Communication Networks Economy

Series Editor
Pierre-Noël Favennec

Communication Networks Economy

Daniel Battu

iSTE

WILEY

First published 2016 in Great Britain and the United States by ISTE Ltd and John Wiley & Sons, Inc.

ISTE Ltd
27-37 St George's Road
London SW19 4EU
UK

www.iste.co.uk

John Wiley & Sons, Inc.
111 River Street
Hoboken, NJ 07030
USA

www.wiley.com

Library of Congress Control Number: 2016948540

British Library Cataloguing-in-Publication Data
A CIP record for this book is available from the British Library
ISBN 978-1-84821-979-3

Contents

Preface

The term *economics* is generally understood to mean *sound management*. This is associated with openness, good faith, accurate figures and integrity in company accounts, with transparency ensured through satisfactory standards of good practice in relation to investors and the clients who have placed their trust in an organization's managers.

Network economics focuses on rational planning, which aims to implement only what is necessary and to provide, in the most effective way possible, universal access to the means of communication best suited to the required purposes. It is clear that this definition largely ignores understandings of what is considered necessary and what could be expensive.

Similarly, a country's economy is the result of an organizational structure in which efficiency is linked to clear objectives. For example, reducing the prices of sought-after industrial goods can encourage the mass export of those products to the international market. For a country's trading to be assessed positively, it should meet the subjective standards that it seeks to achieve. These may include improving the standard of living of its citizens, repaying the national debt, full employment and making culture and leisure available to everyone.

Without establishing strict rules on the maximum rate of debt or its repayment time, comparing the economic situations of nearby countries may be used to justify the trends observed in balance sheets. To make such comparisons, indexes and suitable, stable metrics are necessary, although economic experts continue to disagree strongly on this issue.

On a technical level, the key criteria required by networks as a priority relate to quality of service, confidentiality and communications security. These criteria affect the economies of telecommunications networks, which may themselves be viewed in different ways if each partner has their own opinions:

– the first area concerns the network operator, which must respond to customer demands with the technology at its disposal. With the assistance of manufacturers who define the development and production of equipment, the operator is responsible for its installation and operation throughout the national network, as well as for selling communications services to professional and residential users. The company, while providing services to its clients, must make a profit that allows it to finance future investments and development and research work;

– the second area concerns network users and whether the services and applications offered by operators truly respond to their needs, taking into account the money spent.

Do each of the two parties, operator and customers, receive what they are entitled to from this exchange of services? There are several questions that can be used as examples to illustrate this issue. Is ISDN, with its two available interfaces, properly suited to the requirements of all companies? Why has frame relay undergone such an important development in the American banking sector in particular and why has its operating life been so brief? Has ATM technology, as logical as it appeared to network experts, been an essential improvement to business services? And is the Internet, as popular and rich in applications as it has become, satisfactory to the range of mobile and fixed users in towns and rural areas?

Internet technology offers bidirectional digital connections of varying speeds between users of all forms of connection, be it sound, text or multimedia. For all that, its implementation presents a certain number of economic difficulties for the network operator. Is the operator properly remunerated for its efforts in this area? Are the regulations underpinning the introduction of competition to the communications market wellsuited to managing such complex digital technology?

For almost three centuries, noted authors have posited a number of economic theories. The expression "political economics" ("EcoPo", for French-speaking students, from "économie politique") raises the idea of specific strategic principles or theoretical "natural laws", able to facilitate the management of goods and services production, which itself requires the involvement of political authorities. The economics of telecommunications networks has developed alongside our society. If there are unmistakable connections between real parameters, developments in network technology should now allow them to emerge in a logical, clear and lasting way.

This work does not to attempt to lay the foundations for a theory specifically geared toward the economics of tomorrow's communications networks. On the contrary, this book presents a simplified picture of the principal elements of the economics of a network today, intended to be accessible to ordinary technicians, taking into account experiences they acquire in the field. The major elements associated with network economics that affect the quality of services and the profitability of investments are listed, step by step, accompanied by references to recent economic works. In order to discuss the effect of costs on the suitability of services offered by communications networks, this book is organized into the following five chapters:

– Chapter 1, based on the history of European networks, focuses on aspects related to building a national network.

– Chapter 2 focuses on the structures and architecture of a network.

– Chapter 3 discusses the general themes of regulation and tariff principles, including the latest developments.

– Chapter 4 is dedicated to supply and demand from the perspectives of professional and residential users and network operators.

– Chapter 5 brings together different aspects of the present situations of networks and the effect of the Internet on the economy.

In conclusion, the reader will find an overview of the most significant issues likely to influence the economics of communications networks as they are today.

Daniel BATTU
July 2016

1

The Challenges of Building a Network

The user cost of a network is primarily affected by the guidelines put in place for its construction and management. The quality and security of the network dominate the choices made during its development, as they are indicators of a good level of usage on the part of customers and, consequently, of a good return on investment. However, well before taking the economic aspects of a network into account, conditions in the real world very often make it necessary to establish a network as soon as possible in specific places across the territory for political reasons.

1.1. Construction of a national network

1.1.1. *Network creation priorities*

The construction of a network is the result of will and political strategy. It requires jurisprudence required for managing the so-called "rare resources" attached to network operation. Regulations must be compatible with the development of the network.

1.1.1.1. *Primary physical conditions*

Responding to an unsatisfied demand, communications networks are constructed year after year, although the hypotheses taken into account at the start of planning may be substantially modified over time. The layout of the network should be designed to be able, going forward, to respond to the needs expressed by citizens at the best possible price. Frugality (here meaning the economical use of resources) should be associated with the need to serve the entire population. The layout of the network must correspond to the territory's organizational needs, as decided by its political authorities. In the absence of a specific national directive, the person responsible should act "in the best possible way" and take forward-looking actions. From its inception, the network should be designed to serve the majority of large centers through grids that ensure the duplication of communications transmissions, with the largest number of users able to connect to the national network at a later stage.

The implementation of a national network requires planning made up of various stages and taking place over a number of years. Over time, this plan is often remodeled because of new technologies becoming available, proposals for economic development or new needs expressed by a new distribution of customers in the territory.

1.1.1.2. *Scalability of regulations*

Communications regulations are managed to accommodate any future developments. Initially, the State, which is the source of the project, fixes the legal rules on creating a national operating company (monopoly, cooperative management, administration, delegation to a designated commercial body), as well as the relevant primary legal measures (the rights of public authorities, State requirements, expropriation, certification and status of staff, establishment of user rights and of charges for services open to the public).

The opening of the communications services sector to competition introduces new aspects of regulation, taking into account various standard phases of development (basic services, infrastructure, combining fixed and mobile networks, for example). It is seemingly the relevance of the latest technical and economic events that, as a result of actual demand and economies of scale carried out on network costs, justifies the modification of regulations. The objective is to ensure that the available network services can be accessed by all citizens, in any part of the country and at the best price, without unbalancing the budgets of authorized operating companies, without allowing any of the actors to establish dominance and without calling into question their infrastructure roll out.

1.1.1.3. *Initial investments*

As a result, significant investments only lose value when they are made too quickly and they should correspond to a purpose highly valued by users. A lot of work carried out in preparation for long distance extensions is often unjustified and does not contribute to the repayment of investments.

During the 1980s, network equipment was designed to last approximately 10 years, a period during which the operator could provide a production follow-up, a maintenance service, intellectual property monitoring and an after-sales service. Markets change quickly at present and marketing experts believe that public tastes are organized around a commercial cycle close to 3 years long. Provision of services, especially devices, should therefore be organized around cycles of approximately 3–5 years, subject to revision.

1.1.1.4. *Rare network resources*

Cutting investment costs by reducing some necessary work is feasible, if sharing is possible with other bodies. To facilitate wireless connections, high surfaces and antenna installations for hotspots may be shared. Private companies are often established for this purpose to serve as a single access point to provide competing network operators with fast solutions for the installation of antennas in urban areas. Some roof terraces, selected to be easily converted into relay stations, are accessible in exchange with the building's owners.

Other requirements must be taken into account when creating a network, such as the necessary cable channels, or the need to share the available space with other social amenities (known as common utilities in the United Kingdom), such as the operators of other, public or private, communications networks, highway operation companies, rail networks, quarries or water supply networks.

The frequency ranges for operating wireless connections are available, subject to the authorization of the agent made responsible for those resources by the national regulatory authority. The agent implements the relevant steps adopted internationally by the World Radio Communication Conference (ITU-R) and tenders contracts. Lastly, network addressing (such as TLD), allocated by the managing body, is also a “rare resource” as it is customized through use.

1.1.1.5. *Partnership between operators to serve an area*

Some operators may jointly participate in building and operating a network. Distribution may be organized by geographic area, by splitting the country into zones of influence, or rather by functionality (national and international) or services (telephone, telex and television). The regulatory authority may also decide to open all services to competition among all potential actors in the territory. The tendering of cable installation work to a third party as part of the delegation of public services may be considered to be sharing infrastructure construction.

For example, in 2015, network operators, including Metroweb (which already operates a fiber-optic network in the Milan area), Vodafone and Wind, decided to work together to build a very high speed network in Italy. Operators and service providers can also collaborate to improve the service in a given territory with the agreement of regulatory authorities.

Geosynchronous satellite communications services provide a potential example for collaboration between companies. Following the performances of land-based and submarine fiber-optics, the company Teledesic believes, for its part, that it is possible to balance out the installation costs of satellite networks through the following activities:

– the manufacturing of important microsatellite constellations to connect the entire world to the Internet;

– the launch of these satellites, their provision, their repair and their reconfiguration in orbit to avoid waiting for a replacement;

– the management of ground operations connections.

1.1.2. *Links with the international community*

Every country should be connected to its closest neighbors and those with which it shares a language and culture. These relationships are also essential for trading. However, without being directly connected to every country in the world, a country may ask its nearest neighbors for a common organization to address transit traffic.

Each operator has its own international network of relationships with the operators of its choice. It is clear that competition established nationally between operators must be renegotiated internationally, with the financial aspects of transit best negotiated between partners in the same group of actors, and at the agreed value of the exchange rate between the currencies of the countries concerned (see section 3.7.2.1).

In practical terms, each national operator's network has, for security reasons, several connections dealing with the outflow of international traffic between the dedicated centers in the countries concerned. The international accounts related to this type of traffic are determined on the basis of the average cost of all established international connections, taken in pairs.

1.2. Security of network operations

The security of network operations is based on several considerations:

– network centers should only be accessible to the staff responsible for duties related to maintaining the network;

– the routes of network cables should not show clear vulnerabilities, either to attacks linked to natural phenomena or those caused either on purpose or involuntarily by individuals;

– network equipment should, by its design, respect the rules that allow proper functioning, both electrically and in accordance with the communications protocol used;

– working equipment should be permanently and securely connected to high-reliability power supply equipment;

– warning systems should be prepared to alert maintenance staff to any attempted intrusion or any defect in the equipment.

1.2.1. *Security architecture*

In networks, messages are transmitted through a complex architecture that may use different media (wires, fiber-optic, terrestrial or satellite wireless media) structured to provide duplication of routing. The standardization process has singled out the communications protocols that allow networks to have the highest level of reliability in routing messages, as well as in the confidentiality of their content. The regulations for installing network equipment and the sturdiness of the equipment and its attachments (supply, antennas, instruments for measuring and managing traffic, etc.) ensure the security of this communications network architecture (ITU-T Recommendations series X.800). The architecture for the security of open systems, defined by ITU-T Recommendation X.805, describes the various elements that may be involved depending on the risks incurred. Security is also addressed, considered in relation to measurements, sectors and established plans (Rec.X.810).

1.2.2. *History of security architecture in France*

As an example, in the 1970s, the French communications network rested on structures reorganized in 1945 with the cooperation of the Headquarters of the National Defense. As a matter of priority, the meshed network of cables and radio relay links served the administrative centers (prefectures), with the capital accommodating, through fortified works, the national center for this web of communications. Located in a small Parisian fortification, a transmission exchange was connected to the government quarter, in the 7th *arrondissement* and to political office buildings (such as the Presidential and Prime Ministerial residences). The International Telegraphic Bureau (historically, the primary data center) served news agencies, concentrated north of the *Les Halles* area, close to the printing houses of the major daily newspapers. The information nervous center and its subdivisions, despite their low density, were structured economically and in the interest of maximum security, as per the context of the era. Due to the availability of new technologies (fiber-optics, satellite connections), the number of connections has widely multiplied, so that the security of connections has had to be ensured on the basis of much greater duplication and an increased flexibility, taking into account user mobility. In summary, security architecture must be adapted to the density of the network and to available technologies.

1.2.3. *Operational security of a network*

In all countries, network operators are asked by their governments to participate in fighting crime and terrorism. In Africa, operators who do not succeed in eliminating SIM-free (and therefore untraceable) mobile devices from their network are subject to heavy fines. As in wartime, the current climate of insecurity has tightened the links between different State bodies, which provide access to important information with maximum discretion.

Telecommunications network technicians respond to legal requests for interceptions related to security, defense and counter terrorism. Their activities concern fixed and mobile telephone communications, as well as the Internet. Their duties mostly deal with identifying the people communicating, geolocalizing devices and sometimes blocking Websites glorifying terrorism or featuring child pornography. The legal provisions are at times inadequate and the criminal underworld and its economy are often one step ahead.

It goes without saying that a highly protected network is expensive. A permeable network, on the other hand, attracts large numbers of fraudsters of all kinds, as evidenced by the fact that, in 2015, African networks, serving close to a billion people, suffered an enormous volume of large-scale attacks by cybercriminals, particularly in Côte d'Ivoire, Ghana, Nigeria and Senegal. Since 2013, technical elites dealing with specialized protection have been established in those countries, as well as in Burkina Faso, Niger and Senegal.

1.2.4. *Elements of network security*

Eight major points characterize the different aspects of the security of a communications network (ITU-T. X.series – Data Networks, open systems interconnections and security – X.800 – X.849)[1].

1.2.4.1. *Authentication (Rec. X.811)*

Authentication ensures the validity of the identities claimed by the entities participating in the communication (for example a person, device, service or application). It assures that an entity is not attempting to usurp the identity of another entity (masquerade, masquerade ball, man-in-the-middle) or to reply without authorization to a previous communication made before the breach. The sender and the recipient must both be able to verify the identity of their interlocutor, with the process being more complex in direct communication with visual recognition.

1.2.4.2. *Access control (Rec. X.812)*

Access control (*Know Your Client* [KYC], linked to the e-KYC application) protects against the unauthorized use of network resources. It ensures that only authorized persons or devices can access network elements, information flows, services and applications. A firewall is hardware (made of components and software) responsible for insulating a company's internal network. In particular, firewalls allow network administrators to control access to internal resources from outside (by regulating the flow of traffic to and from those resources).

1.2.4.3. *Data confidentiality (Rec. X.814)*

Only the sender and the target recipient should be able to understand the content of the message sent. This means that the content should be encoded, namely that its data be disguised in such a way that it can then only be understood by an authorized person. Encryption techniques are often based on one or several keys used to

1 http://www.itu.int/dms_pub/itu-t/opb/tut/T-TUT-SEC-2015-TOC-HTM- E.htm.

code the data. These data confidentiality measures protect the data from the risk of disclosure.

1.2.4.4. *Data integrity (Rec. X.815)*

Data integrity ensures the accuracy or completeness of the data sent to the recipient. The data sent are guaranteed to be exempt from any modification, alteration, deletion, creation or reproduction, even if it has undergone processing.

1.2.4.5. *Non-repudiation (Rec. X.813)*

Non-repudiation provides the means to prevent a person or entity from denying having carried out a particular data-related action. It ensures that proof is made available that can then be presented to a third party and used to prove a certain type of event or action has indeed taken place. The digital signature is also based on encryption operations.

1.2.4.6. *Security audit (Rec. X.816)*

The security audit ensures that information is only transmitted between the permitted endpoints and that it is neither diverted not intercepted during its transfer.

1.2.4.7. *Availability and salvage*

The availability of equipment or a system is the percentage of time spent correctly functioning before its first failure added to the total duration when it should have been operational (time spent correctly functioning added to the average repair time). Solutions for data salvage in the event of a crash should be considered among the measures pertaining to availability.

1.2.4.8. *Privacy*

Privacy includes the protection of information that could be inferred by examining a client's activities on the network, that is information about Web sites the user has visited and the user's geographic location, as well as the Internet addresses and names of services used.

1.2.4.9. *Industry participation*

Within the framework of international collaboration established in the "Global Security Agenda", manufacturers, including Trend Micro, have made their expertise and research available to the ITU. They have also submitted reports on threats and security alerts, as well as granting access to their research laboratories. If necessary, alerts are sent out in real time via <ESCAPE>, the ITU's secure portal[2].

These definitions have been adopted by all ITU-T delegates. The application of these definitions to the Internet is at times controversial, with some application providers attempting to introduce definitions based on current regulations in the State of California, for example. However, in such cases international regulations take precedence.

1.3. State communications network

In order to pay the fairest price, be less dependent on network operators and, at the same time, be more secure, French State bodies have an Interministerial State Network (*Réseau Interministériel de l'Etat* (RIE)). Built using the Renater network and constructed in the 1990s, it allows State services to keep functioning in the event of conflict. The RIE interlinks and protects the fixed communications (voice and data) of ministries. A very high speed network architecture allows interministerial information (voice, data and video) to be transmitted without risk of a breach between State sites, as well as a secure connection to the partner networks of ministries via an extranet. It also facilitates the deployment of new applications within ministries. The RIE network's core infrastructure, coordinated by Orange Business Services, is continually supervised by a dedicated and duly authorized team. There are around 10 access points where computer data converges, alongside 17,000 sites connected across the national territory (mainland France and its Overseas Departments and Territories). Since 2016, all French national police force sites have had secure access to the RIE.

---

2 http://www.impact-alliance.org/services/grc-escape.html.

1.4. System for connecting network users

The method of connecting customers to the closest network communications center has changed over time due to the appearance of new technologies. Until 1990, the maximum distance of a wire line between a customer's terminal and the center was 4.5 km (the relation between the length of the necessary cable and the distance as the crow flies is generally close to the square root of 2, here being 3.180 km). This distance has since been between 20 and 100 km according to the method of connection proposed by the operator of the distribution network (2G, 3 and 4G in mobile technology, in hybrid fiber coaxial or in the fiber-optics associated with VDSL2).

Since 2014, due to resources connected to next generation network (NGN) technology, the distance between the customer terminal and the connection center may reach 150 km (network known as the "Tactile Internet").

1.5. The search for regulations and network growth

In the past, economic science has attempted to identify the "natural laws" pertaining to the distribution of goods or the evaluation of a certain number of services, measured by the number of objects or the sum total of transactions, in different-sized towns. The "social mathematics" expected by Condorcet in the 18th Century and the establishment of "rules" or "gold standards" between clear and poorly defined economic concepts have proved difficult over time, with the proposed recurrence models rapidly becoming confronted with exceptions and confusing counterexamples. "Correlation does not imply causation!" Simply because two curves are similar does not mean there is a link between them (sources: BVMI, Federal Statistical Office of Germany – Wikipedia).

1.5.1. *The need for rules in economics*

The rapid economic expansion of Western countries during the 20th Century supported the idea that economic growth could be mathematically linked to manufacturing processes and to the management and correct distribution of the common tools necessary in society. Of course, specific geographic management and a suitable telecommunications framework are necessary to the success of economic activities. Conversely, the telecommunications network must respond to the needs of the national economy. Research has been carried out on these two topics and time has shown that they may sometimes lead to the wrong conclusions and futile investments. Furthermore, developing countries have found themselves damaged by international financial regulations based on imprecise or inappropriate rules.

The changes that have arisen in telecommunications technologies have led industrial companies to manufacture different types of equipment for existing networks. The necessary technical compromises between different generations of hardware enable a response to demand without reducing the quality of service while in fact pursuing the service improvements desired by customers. At present, no single telecommunications technology has succeeded in becoming indispensable in responding to the demands of people around the world. As a result, networks are often made up of equipment from different generations and with different technology. Even in 2016 the Internet is continually evolving, making it difficult to establish workable economic rules to better manage its development.

1.5.2. *Definition of common property*

Each territory has its own features that allow certain geographic areas to accommodate common property earmarked for telecommunications. The land includes specific satellite orbits intended for communications satellites. The high points in the region, as well as some seafront or riverine areas, must be shared between major operators to best position communications channels and devices (radio antennas and underwater connections).

Communications cables, both aerial and underground, may be installed in a straight line through fields or follow roads and standard mountain pass routes. The use of radio frequencies must be subject to resource sharing between the different beneficiaries.

The need to have a national communications network imposes rules that must be observed by citizens. To that effect, the State has key legal prerogatives for implementing a public utility network. More generally, the British refer to "Common Utilities", all public or private organizations (Publicly Owned Utility, Investor-Owned Utility), which compete to set up and manage local, regional or national infrastructure, with the aim of providing all citizens, whether as a monopoly or not, with supplies of gas, electricity and water, communications services, transport, irrigation and waste management.

The choice of the locations and the spatial organization of decision-making centers are decided at the first stage of their establishment and these choices are rarely questioned, as it is difficult to alter the original strategy. National capitals are rarely centrally located. However, in 1100 BC, Xi'an was chosen as the capital of China due to its position in the center of the country, in the province of Shaanxi. In 1520, François I, planned to make Romorantin, a small town close to the center of France, an "ideal city", organized according to popular principles and proposed by Leonardo da Vinci, with two-way traffic and express waterway connections to the rest of Europe, linking central France with the Rhône-Saône axis. Leonardo da Vinci's premature death led to the abandonment of the project in favor of the construction of Chambord castle. History thus decided that Paris would remain the French capital, with Versailles acting as ceremonial residence for official receptions for foreign guests. The location of Paris, both French capital and economic hub, today goes unquestioned. There is no further need to review the economic impact of moving this vast conurbation today, but it must be managed to best suit modern needs, thought by some observers to be quite restrictive [GRA 47].

In countries with planned economies, the concepts of "public needs" and the economy of means to be used have been greatly expanded and connections have been tested in order to attempt to rationalize the distribution of essential goods and services to the populations of urban areas of all sizes. Population movements have even been carried out to make the practices of economic policy correspond with its theoretical requirements. Even today, within the European Community, the concept of optimally sized urban areas, conurbations or regional centers are the subject of debates and economic assessments (see section 1.6.1).

Although nowadays the general consensus is that pursuing the "Ideal City" is nothing short of utopian, the notion remains that there is a physical or mathematical link between society and the technical tools necessary for city life. For the record, several studies have previously been put forward on this subject, the majority of which have been to no avail.

1.5.3. *The first national accounts*

It was François Barrême (1638–1703), a French mathematician, who laid the foundations for national accounts. An expert at the Court of Finances in Paris and an associate of Colbert, he was the inventor of practical accounting tables, known as the universal Barême scale ("*Barême universel*"), widely used to collect royal revenue and control public spending. On the basis of this work, the royal armed services were able to evaluate the number of hay wagons required by the military.

The civil servants with the responsibility for determining whether the State had the material means necessary for its policies and, within various hierarchies, officials such as majordomo François Vatel strove to ensure that everything happened according to schedule. Such pressures remain in our times, whether regarding emergencies related to a presidential visit or the planned opening of the Olympic Games.

To some extent, the suitability of the geographic location, size and practical organization of common goods mechanisms constitutes an indispensable requirement for the correct functioning of a country. The Middle Ages shaped the history and geography of our regions. The establishment of an administrative center may be explained historically through the possibilities offered by coaching inns and by describing the development of stagecoaches. This logic was accepted by all observers who believed that it could be extended to a changing industrial and social world. It is this research on "economic logic", which has led to the construction of the various theoretical models proposed over time.

1.5.4. *Christaller and Lösch's "central place" model*

The theory of Christallerian central places was designed to attempt to explain or justify the size, positioning and number of towns in a region [CHI 33]. Within this framework, towns and villages serve as centers for the distribution of goods and services for a population according to optimization principles based on transport costs. The centers have a supply of goods and services, which they disseminate to users located in their region through a hierarchy of subcenters. Highly used or inexpensive services are offered in small centers close to consumers, just as rarer or more complex services are available in bigger towns, but further apart. The "rational laws" of supply and demand would explain this hierarchy among urban centers, which would be available to families in France of varying sizes. The hexagonal diagrams of centrality in Figure 1.1 are attempts to illustrate a theory that was later refuted by Lösch [LÖS 54].

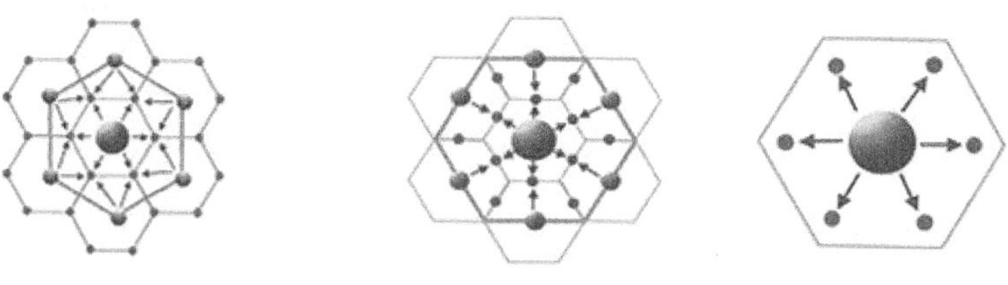

Figure 1.1. *"Christallerian" models of the "centrality" theory (Wikipedia)*

While the principle behind this theory has sometimes been used (land-use planning for the settlement of the polders in the Netherlands, the policy of balanced urban areas in France, the reference model in archeology), it has often failed in rural areas, urban environments and port areas. Pierre Laborde, Senior Lecturer at the University of Bordeaux, has written on this matter that "telex is linked to certain activities (…). But the number of telex terminals per municipality is not always in proportion to its population. Christaller, the original exponent, used telecommunications (the telephone) as a means of expressing geographic centrality. He faced no shortage of criticism and finally renounced the theory himself" [LAB 73]. Despite the criticism, the central places theory is periodically reconsidered, either in relation to the evolution of land-use planning, or the reform of urban centers or the merger of municipalities, with distance (and relative cost) playing a part in the spatial organization of regional activities[3].

1.5.5. *Central places and Pareto*

1.5.5.1. *Regarding central places*

A country's "central places" still tell the story of its history and the human, social and economic transformations that have taken place there. The cantonal framework of France, highlighted by geographer Léon Lalanne in 1863, was established according to large, attractive centers and a six-level regional hierarchical structure [UNI 02]. The economy of territories analyzes the economy of space and distinguishes between urban economies and regional approaches under the growth theory. The distribution proposed at the end of the 19th Century by the Swiss economist and sociologist Vilfredo Pareto, introduced as the "80/20 rule", points out that a small percentage of a population (often) represents the bulk of its value [JUR 60].

1.5.5.2. *Inventory of assets according to Pareto*

In reality, the so-called "Pareto law" is not a "law", strictly speaking, but a method of analysis. Pareto's "log-normal distribution" does not necessarily imply a "magical" 80/20 distribution, but it is a method that can be universally applied by introducing this mathematical instrument to easily separate essential elements from other, less important ones. This dichotomy is viewed as a tool for analysis, particularly of company

3 http://www.hypergeo.eu/article.php3?id_article=13.

management (stock, sales, deliveries, disruptions). In logistics, it means that products with a high value or high turnover should be singled out and subject to tight controls, while products with a lesser value or lower turnover would be under more flexible control. The methodological approach is thereby simplified, even though there is still a gray area in terms of demarcation.

1.5.5.3. *Examples of Pareto distribution*

Pareto's representation demonstrates that, quite often, 20% of the most widely distributed products or the biggest customers constitute 80% of sales revenue, with, conversely, the remaining 80% being 20% of turnover. Applied to French forests in 2010, this representation emphasizes surfaces larger than 4 ha, ignoring the legitimate interests of smallholders.

In this way, in France in 2010, close to 89% of landowners possessed 23.6% of private French forests and 76.3% of the wooded area was made up of plots larger than 4 ha [LAN 10]. The distribution shown in Table 1.1 explains the structure for administrative management of the forests (set up particularly for surfaces greater than 4 ha) and the underrepresentation in forest-holder associations of owners of small wooded areas, who are responsible for 23% of timber production.

Size of wooded plots	No. of owners (thousands)	Percentage of owners	Thousands of hectares sum total	Percentage of surface
Less than 1 ha	2,203	66.5	687	7.5
1–4 ha	729	22.5	1,463	16.1
4–10 ha	231	7	1,425	15.7
10–25 ha	96	3	1,460	16.1
25–100 ha	41	0.8	1,870	20.6
>100 ha	9	0.2	2,174	23.9
Total	3,309	100	9,079	100

Table 1.1. *Distribution of French forest ownership in 2010*

Similarly, the distribution of jobs in American companies in 1970 and 1990 gives rise to the representation shown in Figure 1.2. Over 20 years, despite the 1973/1975 oil crisis, this distribution has remained extremely stable.

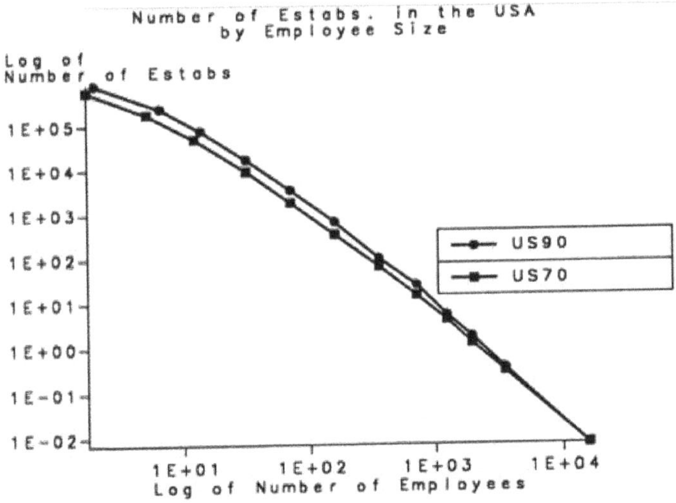

Figure 1.2. *Distribution of jobs in the United States in 1970 and 1990 (County Business Patterns, US Department of Commerce, cited by Paul Tréhin[4])*

1.5.5.4. *Pareto and central places in telecommunications*

The inventory of French telecommunications centers, shown using the Pareto distribution (logarithm of the property of centers, classed in order of importance, on the vertical axis and ranking logarithm on the horizontal axis) gives rise to the diagrams shown in Figure 1.3 [PAU 78].

Above a certain size, the straight line representing the property of the largest centers changes slope in relation to less well-equipped centers (1975). Since 1985, this representation method has shown the existence in the network of three groups of centers, first made up of

4 http://gilles.trehin.urville.pagesperso-orange.fr/distribution_et_loi_de_pareto.htm#_ftnref1.

very large Parisian switching and transmission centers, followed by 20–40 centers in regional capitals and holdings of over a thousand small local centers. It should be noted that this inventory covers the installed equipment's capability to respond to a very high demand for communications without a direct connection to the intensity of actual demand.

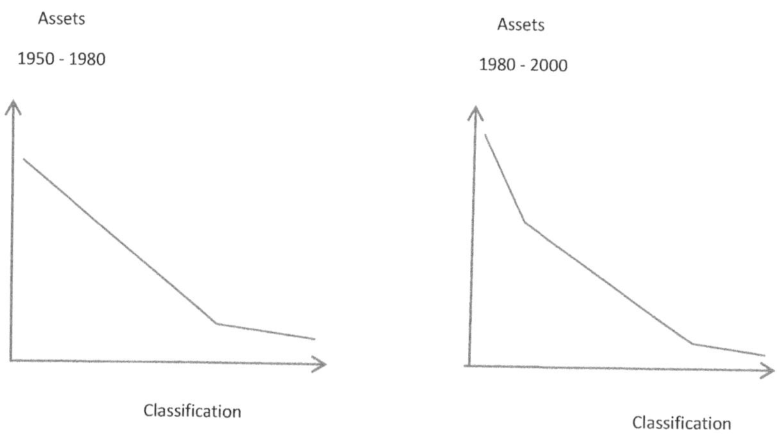

Figure 1.3. *Pareto distribution applied to telecommunications centers*

In practical terms, four remarks can be made on the subject:

– between 1950 and 1980 (left graph), the French network only covered long distances with analog equipment, which had a precise installation modularity. The delay in responding to demand led to large towns being served as a priority and to progressively increasing network capabilities while moving onto medium-sized towns;

– between 1980 and 2000 (right graph), the network had to take into account around a thousand new small centers and introduce numerical type equipment to regional and rural areas. Unfortunately, from an industrial viewpoint, the different groups of materials that had been offered to network operators did not respond to the distribution needs of centers either in terms of equipment modularity or the ability to handle increasing demand for broadband, for example. Network operators have therefore been obliged to manage their equipment with overcapacities in medium-sized centers and undercapacities in large centers;

– since the 2000s, networks have standardized IP packet technology and this has justified the use of high capacity routers in the large centers, and of medium or lower capacity routers;

– as a result of the digitization of networks and the Internet capability that gives each computer system a unique URL address, which can be connected to millions or billions of Internet users, large servers enjoy global connectivity while small personal sites remain continually available, only to receive around a hundred connections per year. This network flexibility has an important technological and economic cost for millions of sites with under-used connections;

– the necessary adaptation of industrial tools to demand can also been seen in freight transport where the use of trucks or large shipping containers is justified by the need to obtain the lowest possible transport costs. Similarly, floating cities serve luxury cruises that last 2–3 weeks for the enjoyment of several thousand tourists on the condition that the cruises visit different port areas around the world in the appropriate seasons.

1.5.6. *Practical logistics rules*

Without having rules supported by network economics, the technicians in charge of building a national network must develop arbitrary rules, which can be modified in accordance with the results on the ground. During the so-called "upgrading" period of building the French network (1960–1975), technical offices relied on the following practice, known as "the tenth". A town of 120,000 inhabitants was considered capable of adding 12,000 subscribers to the telephone network within 5 years, justifying the installation of around 1,200 intercity circuits, composed of 120 multiplex analog devices in 10 primary analog groups along 12 routes.

To address shortfalls, the credit allocated must be used as fairly as possible to avoid any one region being favored over another. This rapid change to a network's potential growth also stabilized the energy needs anticipated for supplying equipment and for the transmission center's desired expansion area. In this way, "the tenth" allowed the standardization of connections between different services (switching, transmission, buildings and power supply) and facilitated network growth (also known as the "Delta LP" program, in reference to the monthly growth in the number of main lines of telephone customers).

1.5.7. *Jipp correlation*

The correlation between economic variables and telephone density provides a reference point for the development status of a country's network. This simple indicator has long been used as a benchmark in official circles, but it only acts as a testament to the political interest shown to investment in technology without further consequence.

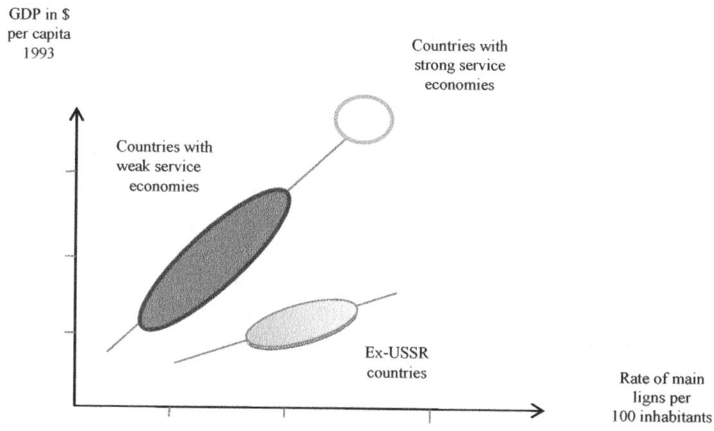

Figure 1.4. *Jipp curve in 1993*

In 1963, a connection was brought to light by Siemens commercial engineer, Professor August Jipp, by comparing the standard of living (GDP in dollars per capita) with the telephone density of the wire network per 100 inhabitants [JIP 63]. The "Jipp curve" is often seen as demonstrating a strong correlation between the growth of telephone density and wealth, with the connection strengthening as wealth increases. Laypersons have sometimes considered this simple relation to be an inviolable law applicable to other networks. In 1993, the telephone density of countries varied in public reporting between a GDP valued at 300 dollars with a line for every 100 inhabitants and a GDP of 20,000 dollars with a line for every inhabitant:

> "Economic development and the development of telecommunications are, in principle, closely linked, although this has yet to be demonstrated. It is indeed the case, looking at an ITU study, covering 164 countries, with a correlation coefficient equal to 0.85, knowing that 1.0 indicates a perfect correlation. The relationship between wealth and telephone teledensity is less pronounced in LDCs (correlation coefficient equal to 0.5), which seems to prove that other factors can influence telecommunications development. Indeed, the line of the "Jipp curve" shows that the most successful countries are those with few inhabitants" (see [OEC 92, LEM 02]).
>
> "If connections actually exist between these variables, there is no proof that these rules are adjustable and introduce reciprocal actions to economics. Other factors, unrelated to information and communications technology (ICT), affect economic development or job creation" [BAT 02].
>
> "This [phenomenon] is often interpreted to mean that there is a causal relationship which works in both directions: better communications generates economic growth and this in turn spurs a need for better communications infrastructure" [KEL 05].
>
> "(The "Jipp curve") "[shows that], if countries were above the curve, they had invested too much and if they were below the curve they had invested too little in their networks. Poorer countries did not have to invest as much as richer countries in telephony; they had to wait until the per capita income increased. Presenting the curve like this does not do justice to Dr. Jipp's insights, but that was how his work was used by policy makers" [OEC XX].

In conclusion, the statistics applied to the social field focus on relative concepts based on numerous conventions. They do not lead to absolute proof being established.

1.6. Land use planning

1.6.1. *Cities, regions and territories*

Theorists in contemporary economics would like to be able to determine the critical mass of population capable of galvanizing a region or territory. According to this process, the possibility of creating businesses and the proliferation of wealth-generating innovations could be linked to defining an approach for a given territory, in conjunction with specific regulations and tax systems. "To promote equal opportunities across the country, France must support the growth of regions which are already favored, where the outlook for productivity is promising" [ECO 15]. The thriving "metropolis", which benefits from an economy of scale, and therefore from relatively low costs and a large workforce, is supposed to share the fruits of its growth and innovations across the nearby territory. Cities are more resistant to economic crises due to their variety of activities. This theoretical logic recommends massive investment in cities in order that they may extend naturally to desertified outlying areas.

The French Planning Commission (*Commissariat au Plan*), led by Jean Monnet from 1946 to 1952, opened up an ideal path for planning and French prospects. In all likelihood, these grand ideas assisted the economic development of the *trente glorieuses* (the "glorious 30" years of economic growth between 1945 and 1975) through the 1963 establishment of DATAR (*délégation interministérielle à l'aménagement du territoire* – Interministerial Delegation for Land-use Planning), and influenced some political motivations up to 1968. Since 1970, the idea of a State presiding over a planning-based Colbertism has seemed ill suited to a European context and a liberal consensus has been considered preferable. Recent studies on land-use planning have shown changes in the concept of "industrial areas", first toward "clusters" and then to "centers of competitivity", without figures or specific examples being presented as incontrovertible and definitive proof.

1.6.2. *Telecommunications buildings*

It is difficult to organize network growth as effectively as possible and to adjust it economically in line with the social sustainability of the associated architecture, namely the areas of buildings dedicated to it. It is said that, at the beginning of the 20th Century, the people responsible for planning and constructing the telephone exchange on the *Rue des Archives* were roundly condemned for having greatly overestimated the needs of manual telephony. The history of the young Parisian telephone network recalls the difficulties faced by the period's policy makers and how, through experience, they discovered the factors for success in the optimal positioning and sizing of the premises required for an urban network [BER 84].

Over the past 60 years, the progressive reduction in the size of electronic equipment has not been foreseeable and, very often, it has not been offset by a large increase in the network's number of customers or its traffic flow. Consequently, in numerous places the buildings dedicated to telecommunications services have proved to be too spacious, as well as unsuited to other uses due to their physical constraints (security, access, ceiling height, space dedicated to back-up power supply). In all countries, it has so far been impossible to reconcile the requirement for telecommunications architecture to be both durable and economically adjustable to needs. Traditional network operators pay the price for this situation because, as owners of large, unsellable and increasingly empty property holdings, they impose an unproductive, but taxable, freezing of investments on their stockholders.

1.6.3. *The search for an optimal model*

France in 2016 is made up of 12 administrative regions, as opposed to 23 previously. These 12 regions cannot be compared with each other or with others in China or the United States. It seems that, globally, there is no connection between high employment rates and area size. On the other hand, other areas are able to remain economically stable without cities. Olivier Grossetti and Olivier Bouba-Olga, respectively, economist and sociologist, believe that there is presently a global trend toward decentralization and that, alongside other complex factors and with the growth in methods of communications being based on the Internet and NGN resources, the trend is probably currently justified. This decentralization of means may also bring about greater operational security [BOU 15].

The economist Frédéric Gilli believes that "all territories may potentially have strategic knowledge and be creators of wealth" [GIL 15]. Accordingly, it is not possible to recommend a single network model with systems for collating customer lines, which are perfectly technologically and economically adapted to each particular urban context. The network operator will consequently be required to build its network carefully and with flexibility in order to respond to urgent and justified demands without losing so much on investments that it might never be profitable. This trend in urban organization, long-recommended without ever being considered a priority in land-use planning, delights network planners.

If there is a practical conclusion to be drawn from the work on "central places", it could concern the design principles for relevant equipment and, in particular, the optimization of modularity, adapted to the sizes of the centers to be served.

Indeed, very often, the fiber-optic capability is well beyond the actual traffic density that a digital subscriber line access multiplexer (DSLAM) can provide, in the same way that, during vacation periods, several buses full of tourists are required to fill an airplane. Conversely, in so-called "off-peak" periods, airplanes take off with empty seats. If today's technology makes wonders possible in the field of communications, it is not possible that, in a democratic society, all customers be compelled to live in towns and subscribe to the same Internet services in order to make shared equipment profitable. Older technology connected a metallic pair to a telephone line and the "Marzinette" system doubled performance by adding another customer to an existing customer's line. By itself, today's technology provides much more and the above-mentioned DSLAM are much better utilized, as they are busier and therefore as profitable in south-eastern regions of China as in Brittany or during the off-season. Just as there are no airplanes in the suburbs, it is not possible to build low-capacity DSLAM with low connection costs. The most advanced technology can only reduce service costs if customers are numerous and active at the same time. As far as equipment is concerned, its actual price does not directly reflect its ability to handle traffic, but functions according to a scale, the first level of which can be expensive.

1.7. So-called "network" laws

1.7.1. *Metcalfe's law*

Metcalfe's law concerns network effects related to information technologies, and therefore also applicable to the Internet or specific social networks. The value of a network depends on its number of users and contributors [$V = N\cdot(N-1)$]. It ensures a critical mass of subscribers.

1.7.2. *Kao's law*

A by-product of Metcalfe's law, Kao's law underlines that the creativity of a network increases exponentially with the diversity and differences of those who devise it. Cultural differences enhance creativity, as well as bringing additional viewpoints.

1.7.3. *Odlyzko's law*

Andrew Odlyzko refutes Metcalfe's law, which he considers too optimistic ($V = n \times \log n$), and introduces the concept of a "critical mass" of users during a network's development.

1.8. Usage metrics for digital technology

1.8.1. *ICT Development Index*

The ICT Development Index (IDI) was recently established by the International Telecommunication Union on the basis of ICT indicators. The intent of this methodology (see Table 1.2) is not to build a law, but it instead justifiably introduces the concept of user metrics. The IDI is a tool adapted for comparative analysis of the most important indicators related to the information society. These standardized metrics are based on 11 indicators, gathered into three groups on access, usage and skills [ITU 14]:

	Index weights (%)	Subindex weights (%)
ICT access 1) Fixed-telephone subscriptions per 100 inhabitants 2) Mobile-cellular telephone subscriptions per 100 inhabitants 3) International Internet bandwidth (bit/sec) per Internet user 4) Percentage of households with a computer 5) Percentage of households with Internet access	0.20 0.20 0.20 0.20 0.20	0.40
ICT usage 6) Percentage of individuals using the Internet 7) Fixed-broadband subscriptions per 100 inhabitants 8) Active mobile-broadband subscriptions per 100 inhabitants	0.33 0.33 0.33	0.40
ICT skills 9) Adult literacy rate 10) Secondary gross enrolment ratio 11) Tertiary gross enrolment ratio	0.33 0.33 0.33	0.20

Table 1.2. *ICT development indexes*

– the ICT access subindex includes five indicators of access (fixed telephony, mobile telephony, international Internet speed, households with computers and households with Internet);

– the intensity of use subindex includes three indicators (Internet users, fixed or wired and mobile broadband);

– the skills subindex includes three indicators (adult literacy, rates of enrolment in secondary and tertiary education). Its weighting is lower in relation to the other two subindexes.

In the ITU's most recent ranking, France placed 17th globally out of 167 countries with an IDI of 8.12 out of 10 [ITU 15].

1.8.2. *Digital barometer*

The General Council for the Economy and the Regulatory Authority for Electronic and Postal Communication have established indicators on ICT usage in France on the basis of interviews with a sample of people representative of the French population (aged 12 and over). This annual survey describes the devices used and their usage by individuals. It focuses on devices, access to fixed and mobile telephony, and to the Internet.

1.9. Conclusions

The examples given here indicate that it is not economics that underpins the initial establishment of a network. The persons responsible make full use of the elements available and share existing resources to carry out implementation at the earliest date possible. Their choices will shape the future directions of network structures. Network extensions must take into account existing structures and at times "forget", in their accounts management, the burden of the heavy investments made. The different theories built around the topic of network construction constitute highly interesting research, but unfortunately they have been unable to find sufficient substantiating circumstances. The reference bases mentioned here in relation to the IDI and the digital barometer provide a better approach to the usage and evolution of digital technologies.

2

Network Structure and Architecture

This chapter provides a general overview of the major elements necessary for establishing the connections which allow applications to be distributed across a network. It also takes the opportunity to recall some conditions essential for providing a good quality of service without requiring excessive expenditure. Aspects relating to the major applications, terminals and network staff are also examined.

2.1. Analog and digital

2.1.1. *Terminology*

The information to be conveyed across networks comes in two forms: analog and digital. The analog form has a continuous signal, even if its amplitude or intensity varies. Digital information, on the other hand, is irregular and is made up of discrete values. The spoken word and music are universally considered to have an analog form. In telecommunications, digital technology uses components that have only two possible electrical states. Transistors perfectly convert binary signals, ones and zeroes, plus signals (+) and minus signals (–) from a telegraph or computer.

2.1.2. *Historical overview*

Until 1970, it was thought that telephone circuits should be used to transmit speech signals with analog technology and that digital technology should be left to data transmission and telegraphy. However, due to the availability of new components, it then became apparent that regional carriers could have an increased number of transmission circuits by implementing digital technology. The progressive improvement of fiber-optic performance strengthened this strategy and transmission gradually converted to digital.

Switching and signaling had already been linked to analog transmission for routing communications along bundles of circuits. The wave of digital technology therefore obliged network operators to reorganize the network's switching and transmission procedures. Several branches of data transmission protocol have been successively tested for conveying digital information in preparation for adopting the International Organization for Standardization model's seven layer organization (Figure 2.1).

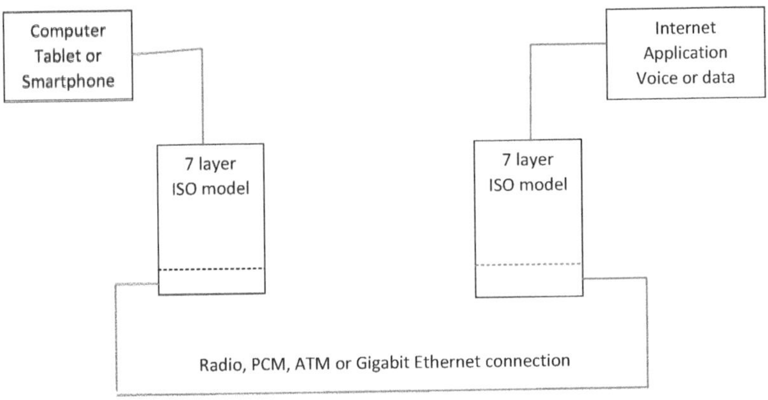

Figure 2.1. *Connections in digital technology*

2.1.3. *Necessary network equipment*

Three types of operational equipment are required for analog and digital networks: equipment for multiplexing connections to make the carriers used as profitable as possible, switching and signaling connections and finally for other long- or close-range line equipment.

Technological development has led to technologies and protocols being selected almost identically around the world, with the aim of profiting from economies of scale and the mass production of components geared toward digital technologies. This allows the connection and distance capabilities of materials to be better utilized, while still offering a good quality of service (reduction of noise levels, in particular).

In wireline digital technology, there are three main types of past protocol regarding information transfer: pulse-code modulation (PCM) that uses a plesiochronous multiplexing hierarchy, asynchronous transfer mode (ATM) and an enhanced Gigabit Ethernet. These three protocols provide a basis for the OSI model's layers (physical layer and link layer), so that the Internet Protocol (IP) protocol in level 3 can benefit from this functional cooperation (Figure 2.1).

Within a national network, there are analog links or digital connections based on different protocols to those referenced above, with adjustments having to be made in order to allow interconnection. On the other hand, fixed-line digital networks are connected to mobile networks through Internet Multimedia Subsystem (IMS), so that the IP protocols may bring the advantages of mobility to all Internet users [BAT 14].

Connecting analog voice terminals to digital terminals is possible on the basis of the adjustments outlined in Figure 2.2. Figure 2.3 shows the connection of fax machines to digital networks through the use of text over IP protocols (ToIP, ITU-T Recommendations, T.37 and T.38).

The IP is the protocol essential to Internet networks, as it allows data packets to be addressed and ensures that they are routed (that is their transit route) to their destination. The IP protocols provide a non-secure transfer for conveying packets. This is why it is most often associated with TCP layer 4 (more secure), and sometimes with the UDP protocol.

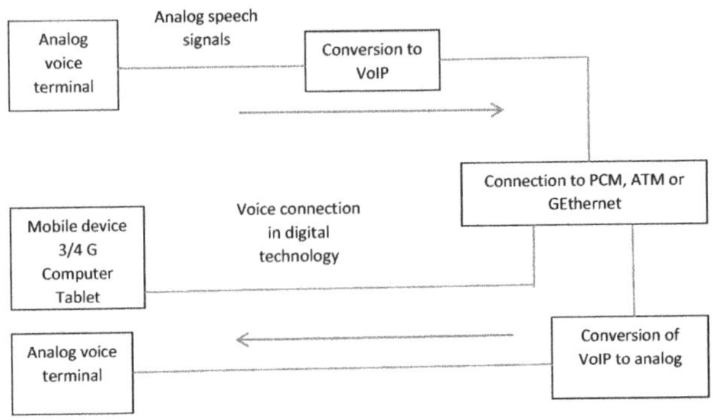

Figure 2.2. *Voice connections on a digital network*

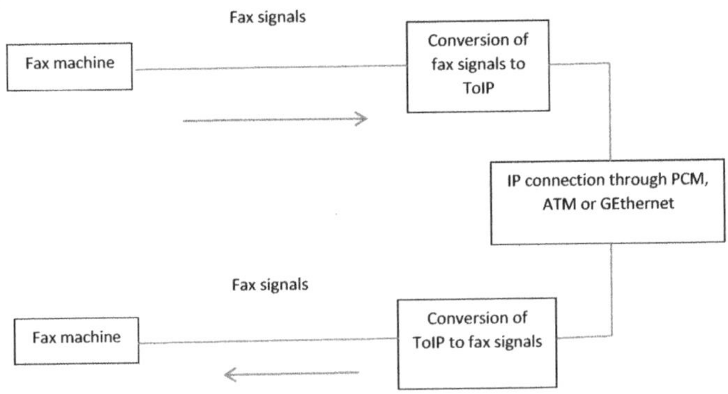

Figure 2.3. *Connection of fax machines to digital networks*

The widespread adoption of standardized TCP/IP or UDP/IP protocols has the advantage of simplifying connection interfaces and reducing implementation costs. Furthermore, the different arrangements shown above are quite similar for both voice and data signals, which contributes to standardizing equipment and facilitating network management. Experience has shown that the major disadvantage of the IP protocol is its simplicity, which facilitates fraud, eavesdropping and identity theft. It is therefore necessary to provide protective procedures for all applications and a particularly thorough maintenance service.

2.1.4. *Traffic metrics*

The movement from analog to digital technology has brought about a fundamental change in the ideas surrounding traffic measurement due to the gradual abandonment of circuit switching in favor of packet switched data transmission.

2.1.4.1. *Analog metrics*

Connecting the caller and the party being called used to require the complete availability of an intercity or international four-wire circuit, where each direction is used less than 50% of the time. An Erlang is defined as the probability that a given circuit is occupied for 1 h and the groups of internationally or regionally connected lines were established according to the likelihood of a call being made during the working day. For example, "a group of 10 analog circuits corresponds to a traffic flow probability of 450 calls during peak time". This idea is still used in circuit-switched voice service networks (public-switched telephone network (PSTN)). With this technology, the manufacturing takes into account the length of calls, the distance between callers and the traffic intensity at the time of connection.

2.1.4.2. *Metrics in digital technology*

All services in fixed-line and mobile digital networks can be considered as IP-based multimedia services (voice over IP (VoIP), messaging service, video, Internet). In this situation, the Erlang laws relating to the probability of voice traffic are not applicable. The flows of data traffic for each of the services under consideration are not symmetrical. For example, a client/server request in HTML protocol produces a highly unbalanced data exchange in a very short time. The key parameters in charging data exchanges focus on the speed of the terminal used, its connection mode (fixed-line or mobile) and the type of subscription taken out with the Internet service provider (ISP). Without having precise metrics to assess the traffic across all available services, the network operator prepares specialized routing channels that are designed to be larger than the maximum amount of traffic predicted.

While digital technology is much more economical than analog in terms of transportation, the need to maintain traffic flow is always achieved through secure equipment, and more than the required amount must therefore be invested in all services proposed.

2.2. Distribution network

2.2.1. *Wholesale and retail markets*

The distribution network is made up of all subscriber lines connected to the mainframe. The operators aim to connect all subscribers and offer them the best possible digital bandwidth. Since 2003, ITU, at the request of GATS, has established two communications markets: the services and applications retail market targeted at the general public (business to consumers (B2C)) and companies (business to business (B2B)) and the wholesale market for owners and operators carrying out bulk resale or using these resources for their own network (Table 2.1 and Figure 2.4).

The requirement for "wholesale subscriptions" was introduced to France in 2005 in the form of regulations linked to supporting new operators and competitors of established operators and unbundling measures for subscriber connections (leased to new operators).

Markets	Actors
Wholesale market	Manager of connected objects Public initiative networks Dark fiber operators Mobile virtual network aggregator
Mixed markets	Managers of fixed-line and mobile multimarket networks International operators Software and computing services companies
Retail market	Fixed-line network manager – General public Virtual mobile network manager – General public Mobile network manager – General public Virtual mobile network manager – Business Fixed network manager – Business

Table 2.1. *Division of the areas of managers and operators*

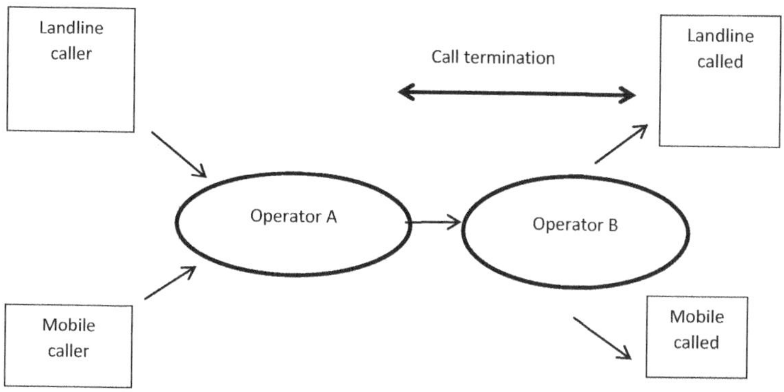

Figure 2.4. *Call termination*

The wholesale market of managers and operators concerns B2B, B2C and carrier to carrier services and applications. As an indication, in 2015, almost 1,700 operators were active and registered with the French communications regulatory authority (ARCEP). There are around 100 public initiative networks in France, which are often the result of local authorities wishing to operate their own networks with the aim of eventually serving 43% of the French population.

The competition between managers and operators does not preclude cooperation between wholesale partners by pooling mobile network equipment or hosting unlicensed mobile virtual network operators. The wholesale market covers refile traffic access to interconnection equipment and renting equipment to fixed-line and mobile networks. Unbundling is a regulatory procedure that allows competitors of the incumbent operators to use the copper pairs of customer lines to their advantage in return for a specific cost-oriented fee. Throughout the national territory, different operators interconnect their networks and bill each other at the agreed rate for the call terminations made (see section 3.5).

An Over-The-Top (OTT) service (see Glossary) is defined as a service that, using the Internet, circumvents the channels of traditional telecommunications network operators without asking for their consent. OTT services are making progress, especially in the mobile sector. Under the Net Neutrality principle, and while positively viewing the significant contributions provided by OTT services to their customers, network operators would like the bit rate transfers made through the channels that they have built and maintain to lead to fair payment.

"Call termination" is a term used when billing communications between managers and operators. It relates to the last part of the service or application provided by a partner network. Due to changes to the various implementation practices that have historically been used (see section 2.2.2), different subscriber connection structures must be taken into consideration in distribution networks.

The different systems mentioned in Figure 2.5 allow the data flows to be gathered at the head end of the network in PCM, ATM or Gigabit Ethernet.

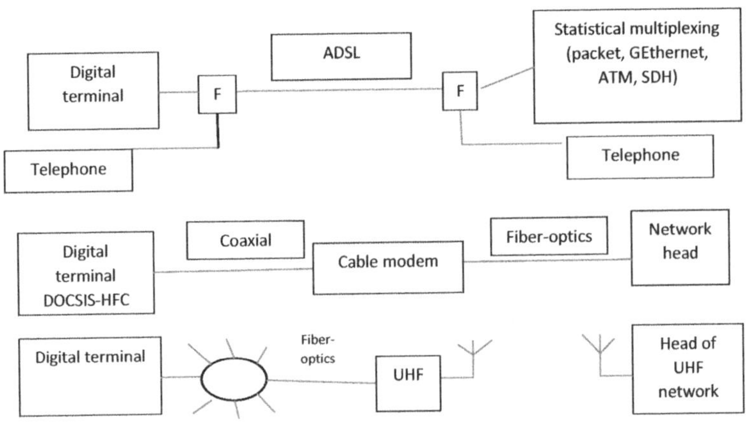

Figure 2.5. *Subscriber connections in ADSL, HFC and radio-cable composite*

2.2.2. *The different structures of fixed subscriber lines*

2.2.2.1. *The analog subscriber line*

The analog subscriber line is represented by a two-wire metal line in a buried or standalone overhead cable that connects the subscriber's telephone terminal to the telephone exchange's connecting device. It transmits, through a continuous direct current, voice, fax or modem – called telephone – signals associated with the signaling relevant to those services. In general, this subscriber line has no active equipment and does not exceed 4.5 or 8 km, depending on the components used.

From the creation of the French telephone network until 1975, the installation of analog subscriber lines by the PTT administrations – who had a monopoly on this activity – was carried out upon request, unit by unit, subscriber by subscriber. For remote subscribers, this connection was provided upon payment by an individual, calculated according to the distance separating the customer from the edge of the conurbation. Between 1975 and 1980, bundles of subscriber lines were systematically installed, neighborhood by neighborhood and region by region in order to offer communication services to everyone, without exception. A telephone pair was available in front of every apartment and every house, and ready to connect to the telephone exchange upon signature of a commercial contract.

2.2.2.2. *Asymmetric digital subscriber line*

Connected to filtering devices at each end of the line, this link concurrently provides lower frequency (0–4 kHz) analog telephony and bidirectional transfers of digital signals at higher frequencies (between 180 and 20 MHz) for DSL speeds between 500 kbps and 1 Gbps, depending on the distance. This economical use of subscriber line carriers has only been made possible after long negotiations on the international standards to be adopted in order to mass produce components for the global market. It was an audacious gamble as, in all countries, the "distribution" part of the network has been organized without international dialog. Network operators have therefore only discovered the problems associated with rolling out the asymmetric digital subscriber line (ADSL) very gradually and, in response, manufacturers have improved protocols pertaining to digital management of signals on subscriber lines in both directions of transmission.

2.2.2.3. *The digital subscriber line*

Made up of several sections according to capacity, a digital subscriber line can today reach 150–200 km. Various connection techniques are favored by network operators, with the economic aspects of each being linked the number of subscribers collected at breaking point as well as to the traffic needs of each.

Permanent radio links and fiber-optics allow subscribers to be connected to their assigned center (see Figures 2.5–2.7). Combinations of carriers are possible on subscriber lines. For example, the fiber can be extended by a copper pair (in very high bit rate DSL (VDSL)), a coaxial cable (hybrid fiber-coaxial (HFC)) or a radio link. The speed is therefore not limited by distance. The choice of speed is determined by the local economic conditions and the extent of demand in the area where the fiber ends, with the operator making optimal use of the existing resources in the client's neighborhood.

Fiber is not currently necessary in France, but experts are convinced that by 2020 it will be used "for the Internet, as for television". The state, which has already committed funds, decided that, "in order to be ready, preparation must begin now". The €20 billion needed to roll out broadband will be provided by private operators and the state. Networks will be rented by operators and will bring money to communities, which should cover half of the costs incurred by the State, with the balance being paid by taxpayers.

Distinctions must be made between:

– *FTTB*: fiber to the building;

– *FTTC*: fiber to the curb;

– *FTTCab*: fiber to the cab;

– *FTTN*: fiber to the node;

– *FTTH/FTTP*: fiber to the home/premises.

In the United Kingdom, British Telecom committed to connecting 2,000 residents and business in the Huntingdon region in 2015 to G.Fast 330 Mbps technology, a speed that will be raised to 500 Mbps in 2016 (ITU-T Recommendations G.9700 and G.9701). The fiber is connected to two copper pairs whose length must remain less than 100 or 500 m (FTTB). The entire British broadband plan (Broadband Delivery UK) is an investment of over a billion pounds sterling.

According to IDATE, the growth outlook for broadband subscriptions remains very strong globally, as they represented only 37% of all subscriptions at the end of 2014 (in comparison with 25% in 2013) and are likely to double again in 2017.

FTTH/B affected 62% of FTTx subscribers in 2014. The growth of FTTH/B subscribers will remain very positive up to 2019. This technology is widely used in Asia, Africa and Latin America.

VDSL represented only 11% of FTTx subscribers in 2014, as in 2013, but it has great potential in Europe to maximize investments in copper networks.

In the United States, for the 2015–2018 period, Comcast is developing a cable network in accordance with HFC standards, named "DOCSIS 3.1", with the technology guaranteeing download speeds of 10 Gbps and upload speeds of 1 Gbps (Figure 2.6).

Using a single installed optical fiber, passive optic networks (PON) topology allows a large number of subscribers to be served. Various solutions are available to make PONs profitable depending on distance and the number of customers served. During the European Conference on Optical Communications (ECOC) event, participating manufacturers and operators presented their solutions on the profitability of investments that had already been made to provide customers with broadband connections. A frequently discussed objective concerns how to provide 1 Gbps as economically as possible for N subscribers in such a way as to be compatible with 4 K and 16 K television standards. This is connected to the growing demand for fiber to the home, or FTTH, as well as to the interconnection requirements of data centers. Figure 2.7 shows a G-EPON system at 10 Gbps shared via Ethernet for passive optic networks (without power supply equipment to predict the volume of connections).

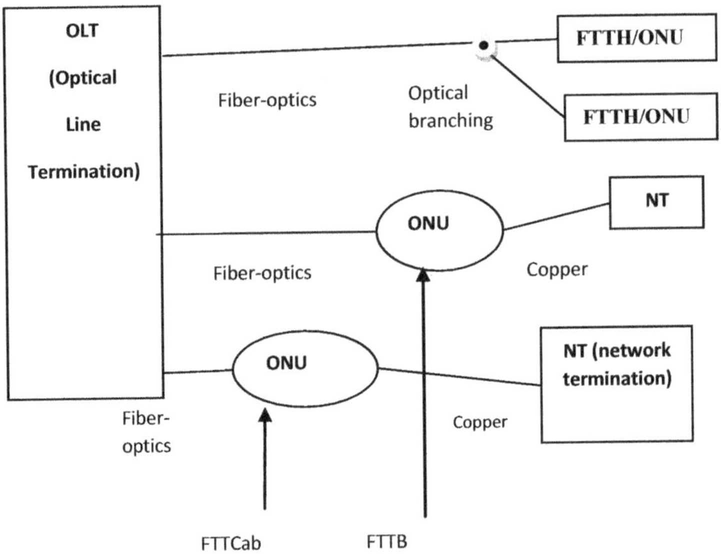

Figure 2.6. *FTTx arrangements*

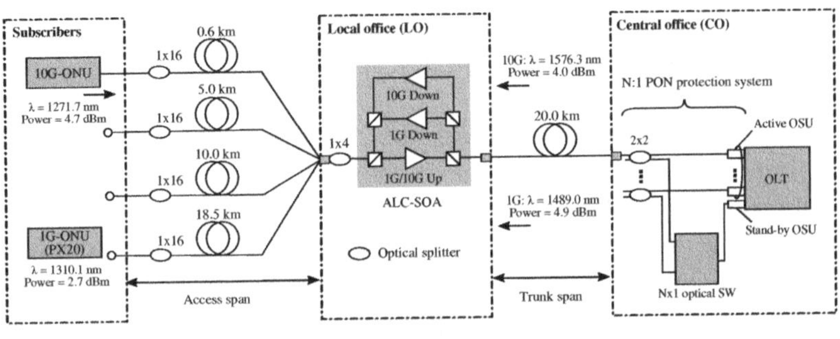

Figure 2.7. *10G-EPON system for long distances and coupling across an access network infrastructure (source: ECOC Cannes 2014, www.photoniques.com)*

Network operators specialize in providing end-to-end fiber-optic connections for businesses, which demand broadband with a high degree of availability.

2.2.2.4. *Fiber versus copper*

It is clear that, in distribution networks, fiber and copper will coexist to bring broadband services to users. However, the "copper switch-off", which would allow savings to be made in network maintenance tasks, remains controversial. On a practical level, the need to manage the transition carefully in France seems to be accepted. In the United Kingdom and Australia, operators who were in favor of fiber seem to have lost their enthusiasm for it, taking into account the related costs recorded in areas with scattered populations. They agree with the opinions of Chinese decision makers, who are in favor of FTTC coupled with VDSL2, and American operators, who often use the HFC system. Apart from the costs resulting from implementing these broadband connections, it should be recalled that terminals connected to fiber-optics still consume large amounts of electricity, even though progress has been made recently in the field of lasers. It appears that the encouragement to convert access networks to fiber in order to speed up broadband provision did not take into account the project's parameters. There are other, less expensive means of bringing broadband to remote locations. The fulfillment of the United Kingdom's "Digital Britain" plan, on the other hand, seems to show that not all citizens are ready to accept the entire new range of services. And what are these fast broadband services?

2.2.2.5. *Connection to power line communication technology*

Power-line communication technology is intended to open up new uses, such as renewable energy production (smart grid) or new needs linked to charging electric vehicles. Households producing renewable energy should contribute to better real-time management of the electricity network within the context of the energy transition. Digital data on production or electricity demand is transmitted using low and medium voltage connections from the transformer to the subscriber's meter. New equipment should be developed for these new uses.

2.2.3. *Connections via radio link*

Operating a radio network requires authorization from the relevant recipient, with the participation of the French regulatory authority ARCEP, to use frequencies, spectrum planning and regulatory procedures.

2.2.3.1. *Overview*

The GSM (or 2G mobile system) provides a connection to fixed-line and mobile networks for speech and narrowband data transmission, as well as a text messaging system. Some countries intend to abandon this technology, which is considered less effective in terms of frequency use.

The third, fourth and fifth generations of mobile telephony have improved previous services through faster connection to the Internet. Speech is transmitted in packet mode. Systems can have complementary or competing pairs of solutions on certain technical and economic aspects, such as Wi-Fi and 5G, or ZigBee and Bluetooth. Table 2.2 shows the uses of these communications services, which utilize one or two cascade systems if required.

Generation	Voice	Fax	Messaging	Video	Speed	Obs.
2G	Yes	Yes	SMS	No	40–500 kbps	In decline
3G	Yes	Yes	SMS and Internet	Yes	200 kbps–28 Mbps	In use
4G	Yes	Yes	SMS, IoT and Internet	Yes	100 Mbps–1 Gbps	In use
5G	Yes	Yes	SMS, IoT and Internet	Yes	100 × 4G	By 2020
Wi-Fi	Yes	Yes	Internet	Yes	6–54 Mbps	Multiple standards
Bluetooth	Yes (profile)	Yes (profile)	Short-range data (profile)	Yes	750 kbps to 2.1 Mbps	Short distances
Zigbee	No	No	Short-range data	No	20–250 kbps	Economical
LoRa LoRaWAN	No	No	Short-range radio data	Yes	0.3–50 kbps Being investigated for IoT	Low power

Low Power Wide Area Networks (LPWAN) being investigated.

Table 2.2. *Major global wireless communications systems*

For around 20 years, changes to these new wireless communications systems have increased the variety of performances in terms of the speed and range of service, and supplementary services. The value of a useful data rate is even more important than the size of the available frequency range as it is located in an operating spectrum navigating urban obstacles. Modulation techniques are chosen by operators on the basis of the quality of real service on the ground and the price level of their implementation.

The two transmission directions of a communication can share the same modulation channel or instead be placed in separate channels. For economic reasons, time division multiple access and code division multiple access are the most used, albeit with certain limitations. To conserve frequency resources, some systems use time division duplexing (TDD) or frequency division duplexing (FDD) duplex techniques:

1) time division duplexing: a single frequency channel is used for both directions of transmission. TDD is also used for point-to-multipoint communications. It presupposes on-call time to change the operational direction of the transmission. TDD is suited to unpredictable asymmetric flows;

2) frequency division duplexing: a frequency channel is reserved for each of the directions of transmission. FDD is suited to point-to-point communications where the flow is symmetrical and predictable.

Multiple input multiple output (MIMO) technology combines a network of antennas and a diverse multipath propagation during transmission and reception to increase speeds. The signal to be transmitted is split into different signals, which are transmitted according to an algorithm agreed by several transmitters and received by several receivers. Conceived in 1983, the MIMO technique is currently used in variants of spatial diversity, in spatial multiplexing and in antenna networks (technology known as "Beamforming"). Over time, operators have informed manufacturers the advantages and disadvantages of the different techniques used to connect subscribers, as well as the potential adjustments, which explains the multitude of solutions currently available.

4G/LTE technology is widely used around the world (451 operational networks in 151 countries), particularly in FDD, and in 2016 it is available to almost 88% of mobile network users. The most recent developments using groups of carriers allow peak data rates to reach 150 Mbps in the downlink direction and 300 Mbps in the uplink direction. The number of LTE users should surpass the number of subscribers to 3G technology in 2020. The increase in the number of terminations is driving the use of IPv6 addressing, even on Wi-Fi. 5G standards will soon be published by the ITU using the reference IMT-2020.

4G generates large revenues due to the capabilities of its terminals and the availability of mobile broadband. On average, a 4G subscriber consumes an amount of data two to three times greater than when using 3G. The problem presented by total coverage of a country is that less densely populated areas, which amount for 18% of the population and 63% of the country, are still sparsely covered. At the end of 2015, the French population received 4G coverage within a range of 60–75% according to operators. It will be between 95–99% at the end of 2018.

On the whole, the radio spectrum is underused, be it due to an imbalance between supply and demand, or the need for every operator to build up a reserve. On the other hand, other wireless methods of access have been successfully tested (shared access with or without authorization, dynamic access using cognitive radio). This situation is likely to slightly reduce the costs of issuing licenses. The number of unique mobile subscribers around the world was 4.7 billion at the end of 2015.

2.2.3.2. *The issue of cellular radio technologies*

4G cellular networks are able to offer theoretical data rates greater than 100 Mbps, or even greater than 1 Gbps (standards defined by the ITU within the framework of IMT-Advanced specifications). In practice, the data rates are in the range of a few dozen Mbps, since the data rate is shared between active terminals within the relevant cell.

The real data rate may be less than 10% of the theoretical data rate. On the other hand, due to the introduction of new mobile terminals (smartphones and tablets), users need higher data rates to be able to benefit from the quality and interactive content promised. To stream a high-quality video, the company Netflix recommends a rate of at least 5 Mbps, but in fact streaming requires a higher data rate. Applications make an ordinary cellular network vulnerable. Networks in large American towns experience periods of traffic congestion and speed restrictions. Due to spectrum limitations and cellular technology's own constraints, it is impossible to exaggerate the size of the wireless network as happens in communications channels based on fiber-optics. In other words, wireless technology is not always able to resolve the conflict between supply and demand, even over the long term. On the other hand, recent economic studies have tried to identify the appropriate strategic moment for a network operating a 3G network to consider introducing a 4G network in a world open to competition [HUA 13].

In Africa, it has emerged that the cost of the radio spectrum is not necessarily linked to the quantity of available frequencies, the size of the population and the revenue estimated to come in from the communications market. According to a study carried out in 2015 by the consultancy firm IDC, several network operators have had to reconsider their plans for development in Africa due to the low level of infrastructure, particularly in electricity supply. They must concentrate on developing products and services that appeal directly to the needs of the local business market and to small- and medium-sized businesses in particular.

In its early stages, in 2012 and 2013, VoLTE technology (the voice over technology of 4G/LTE mobile telephony) quite discouraged operators, as it was expensive and required the implementation of the IMS signaling platform to ensure the transmission of speech along packet-switched circuits and vice versa. Many operators have considered it preferable (mistakenly, it seems!) to wait for the volume of voice traffic to be transferred, believing it more important, and that a high demand for Internet services is becoming apparent as other, more conventional solutions were available [BAT 14].

A new Wi-Fi technology, christened "passive Wi-Fi", using 10,000 times less energy than existing technology, is being tested. Besides the interest in the useful life of the batteries of mobile terminals, "passive Wi-Fi" will bring to the Internet of Things (IoT) a range of smaller, lighter and longer lasting sensors and transmitters. In a nutshell, when there are multiple ways of supplying radio connections, it falls to operators to decide which is most appropriate for their market, considered within the parameters of modulation, purchasing costs, equipment maintenance costs, coverage and capacity in the number of active subscribers per km^2, per hour and per day. Open source usage and the use of software defined radio are often economical solutions for adapting different systems (3, 4 and 5G).

The numerous Wi-Fi standards established by the American company IEEE define radio communications networks in office spaces and urban areas. These Wi-Fi networks are often used in addition to 3 and 4G cellular networks and, in a recent development, their implementation may be proposed by third-party organizations (OTT in India and the United States, for example), together with network operators.

For operators who experience the 3G market on a daily basis, it is not always necessary to invest in 4G, knowing that customers may not show a great response to the offer. In addition, the coverage obligation linked to these 4G licenses weigh quite heavily on investments. Customers satisfied with 3G services do not necessarily have a need for the stunning, but expensive, benefits of 4G.

2.2.3.3. *Radio connection via satellite*

Connection of customers can also be carried out by satellite radio link or a stable platform (high altitude platform systems (HAPS)) located at a high altitude (last diagram in Figure 2.5).

Several satellite services companies, including Globalstar, Inmarsat, Iridium, Thuraya and United World Telecom, market their fixed-line or mobile digital services to companies worldwide. For example, as part of its BGAN offer, Inmarsat provides channels at speeds of 432 kbps per channel.

HAPS located at altitudes of around 6–18 km are able to establish connections in the band of 10–48 GHz for areas 200 km in diameter and user data rates between 48 kbps and 155 Mbps. Several projects have been announced in this area. Facebook is planning to launch a solar-powered drone, named “Aquila”, in order to bring the Internet to the most remote parts of the world. Google has tested various solutions, including hot-air balloons (Project “Loon”).

To bring the Internet to everyone, manufacturers of satellites and launchers are proposing satellite constellations connected to fixed-line and mobile telecommunications networks. On the one hand, Virgin intends to put OneWeb, a fleet of 648 satellites, into orbit between now and 2018. On the other hand, SpaceX and Google want to launch almost 4,000 lightweight and inexpensive satellites into low orbit (1,200 km). Lastly, satellite operator O3b (the “Other 3 billion” customers) is about to complete the deployment of a constellation of 16 satellites.

Eutelsat Communications and Facebook are planning to launch low altitude satellites to allow all African users to access the Internet. The payload of the AMOS-6 satellite could be used in this way for several bundles offering coverage of western, eastern and southern Africa from the end of 2016.

A permanent high-speed connection is at the center of large satellite constellation projects for both businesses and families. Future applications will respond to real and specific demands, whether they are for architectural firms, land-use services, hydrography or family needs. In such cases, technology no longer drives research, and it is users who will develop the most commercially relevant systems. As regards quality of service, it should be noted that satellite communications can be silently listened to by anyone, that they can be scrambled deliberately and finally that parts of radio frequencies can sometimes be used without permission.

2.2.3.4. *High-speed mobile subscribers*

The E3Network project, financed by the European Union, consists of creating a transceiver in the millimeter band (E-band, that is 71–76 GHz and 81–86 GHz) in order to be able to offer speeds of 80 Gbps for the short connecting links (~1 km) needed for future networks.

On the assumption of a very high demand for mobile broadband in the next 10–15 years, additional frequencies will be required and studies are being carried out to consider the use of the "C-band" (3.4–4.2 GHz range), which is particularly suited to changes to 4G and the future 5G. The uses of the "C-band" vary from one country to another, with some wholly or partially reserving it for satellite applications. Another possibility could be to increase the speeds of Wi-Fi hotspots.

2.2.3.5. *Airplane Wi-Fi*

For reasons related to commercial competition, airlines catering to tourists intend to install Wi-Fi hotspots on long- and medium-haul carriers. The bilateral connection to the terrestrial Internet network will be guaranteed through a satellite link or through terrestrial mobile telephony relays. Cultural or sport-related information services will thereby be arranged throughout the plane, as well as teleshopping services and voice and text communications. The cost of equipping a plane, to the tune of US $80,000–400 000, depending on the continents flown over, could be quickly paid off, even if usage only reaches 6% of passengers taking flights under 2 h long. The economic calculations here take account of the ensuing increase in the prestige and image of the airline's brand. However, even if the technology has a low profitability, the availability of the applications has become essential to some customers.

2.2.3.6. *The effects of radio waves on health*

There are Wi-Fi distribution systems and 3 and 4G systems that consolidate bundles into 2.4 and 5 GHz bands, while remaining economical to purchase and run. Reducing the power of radio transmitters has become a major concern, especially as the majority of radio networks transmit around towns and in school areas.

Several studies are being carried out in a number of European laboratories regarding the biological effects of waves emitted by radio equipment. In 2016, WHO intends to proceed with a formal risk evaluation of all the health effects linked to exposure to radio frequency fields (see Appendix 2).

2.3. Long-distance connections

In 2015, long-distance circuits were constructed on fiber-optics, both on land and underwater. Fiber-optics offer the best quality of transmission with a very short transmission time. The equipment installed using cable connections provide an important time delay linked to the presence of filtration systems. The use of communications satellites is also disadvantaged by the transmission delay, which is connected to the distance to be covered between transmitter and receiver.

The maze of optical links usable in 2016 is very dense and it most often ensures routing security for the main strands of international traffic. At a national level, network operators have been assigned the task of providing all towns with fiber-optics and they have been encouraged to share their resources.

The significant increase in international traffic has led to the traffic capabilities of submarine circuits to multiply by several thousand. These cables transported 99% of global Internet traffic in 2015, with the balance being taken care of by direct cable or satellite links. Four leading businesses ensure the laying and maintenance of submarine cables worldwide: Alcatel-Lucent Marine Networks, TE-Subcom, NEC and Fujitsu. International consortiums share the financing of these connections with international financial institutions, supported by large network operators. It seems that the overcapacity observed in the North Atlantic, caused by excesses linked to the 1990s dot-com bubble, will be soon be able to be reduced.

Participating in the investment of a submarine connection is less expensive than renting circuits. A submarine cable project is an investment of some several hundred million dollars with a production timeframe of 1–3 years, depending on size, and an expected operating life of 25 years (the current operating life of satellites is limited to 12 years). In 2015, Asia became the most active area and submarine technology must remain ahead of demand in order to avoid restricting the development of applications and traffic [ARC 12b].

The Orange Group stands out due to its great involvement in the design, construction and operation of submarine cables. Its subsidiary, Orange Marine, manages six cable ships from its marine bases or foreign ports for laying and maintaining underwater cables. As the sole or partial owner of several submarine cable systems, the Group ranks among the world's biggest owners of submarine connections. These submarine cables transport almost all international communications. They are the ideal means for connecting continents and isolated countries. The economic concerns around laying and maintaining submarine cables have become highly strategic.

2.4. Next-generation networks

2.4.1. *Next-generation network structure*

Next-generation networks (NGNs) connect the Internet to digital switched networks. Under construction in much of the world, they are gradually replacing digital telephone networks. They use wired and radio carriers to transport digital information using the IP. The links between different Internet networks are managed by ISOC (Internet Society). They lead to IXP/GIX (peer-to-peer (P2P)) platforms, which are interconnected. The interface of the subscriber's equipment on an NGN is known as next-generation access. The interface between circuit-switched networks and packet networks is ensured through a signaling platform called "IMS". The role of IMS is also to connect fixed IP networks to mobile IP networks (see Figure 2.8).

In this way, a mobile or fixed terminal can be connected to a voice terminal or a Uniform Resource Locator (URL) address on the Internet network to take advantage of a data application. With the Internet currently being a borderless network, all network users can log on to any information database in the world.

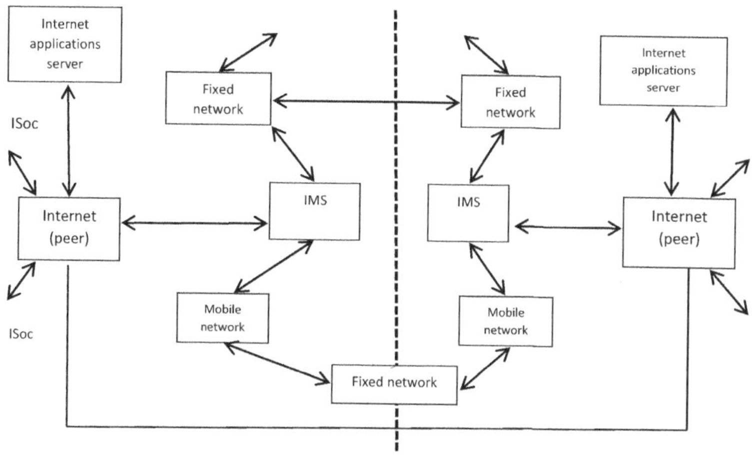

Figure 2.8. *The concept of connections between the Internet and the fixed and mobile networks of two countries. The dotted vertical line represents the border between two neighboring countries. (ISOC, Internet Society – IMS Internet Multimedia Subsystem – Peer: Pair or GIX). (Source: Migration scenarios from legacy networks to NGN in developing countries (ITU-2013))*

Figure 2.9 shows an image of multiservice provision for subscribers. From the center of the network (right side of the figure), around 100 sites receive fiber-optic service and subscribers (left side of the figure), split into 5,500 groupings, enjoy applications carried by ADSL connection, radio link, or fiber and coaxial link-ups.

NGN have four significant advantages as follows:

– the cost price of a packet network is very low (the switching point costs €1 per customer instead of €3,000 in circuit switching);

– the centers have a broader geographic coverage (300 km in diameter instead of 8 km for an analog network);

– fixed and mobile terminals can communicate with each other and access Internet applications;

– and the global distribution of URL addresses gives application platforms a universal opening in IP protocols.

However, digital technology consumes a lot of energy. It appears quite vulnerable to electric shocks, as well as to incidents caused by hackers.

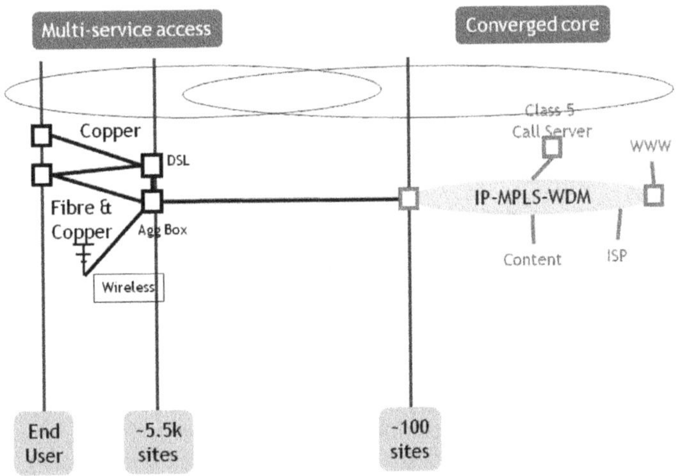

Figure 2.9. *Structure of customer collection in NGN (Source: ITU)*

2.4.2. *Energy supply*

Telecommunications centers have their own secure, double-circuit supply system that guarantees them several hours of autonomy. Some connections require the implementation of remote supply systems, which use metallic or coaxial pairs. A certain number of centers can be powered by solar or wind energy. Recently, supply systems have been developed that combine photovoltaic energy resources and their storage using lithium ion batteries with an operating life of 10,000 cycles (meaning an operating life of around 20 years, similar to that of photovoltaic panels), which are therefore highly profitable when compared with the operating life of a public network (Appendix 3).

2.4.3. *The relationship between traffic and time*

The traffic that flows through network equipment and arteries is simply a reflection of user activity. In urban centers, the activity of professional users is added to residential activity. The scarcity of intercity resources has, in the past, led to restrictions being imposed on the traffic flows through tariff modulations. Appendices 4 and 5, respectively, show the variation in telephone traffic times in 1975 and the tariff modulations proposed to the public in 1958 for intercity connections.

Today, the volume of fixed-line and mobile telephone traffic only shows a reduction around lunchtime, and the growth in traffic is constant between 6 and 10 a.m, proceeding to a slow decline from 8 or 9 p.m. depending on the location and the season.

It should be recalled that voice communications traffic over long distances leads to shorter conversations than over short distances. In this context, the so-called "time zone" effect is particularly marked. Beyond three time zones, the number of possible contact hours dwindles, making groups of circuits particularly expensive due to their low usage. Very fortunately, digital technology allows image and sound to be connected to other data streams, which explains the large reduction in international connection tariffs and applications offered by OTT services (including offers from Skype and social networks) being partially free to use.

2.5. Internet

2.5.1. *Types of Internet connection*

Four types of Internet connection can be identified: fixed Internet, mobile Internet, the IoT and Tactile Internet.

2.5.1.1. *Fixed Internet*

Fixed Internet allows users to exchange voice services, data and multimedia content at the fastest speeds. Connecting customer devices to the network is carried out through a wired carrier using copper, fiber-optics or a combination of the two mediums.

2.5.1.2. *Mobile Internet*

Mobile Internet connects people everywhere via radio link with 2, 3 and 4G technologies, or via Wi-Fi networks, with speeds able to reach 1 Gbps. Mobile Internet is connected to fixed networks through the IMS signaling system. Mobile Internet is suitable for exchanging e-mails, pictures and videos in near real time. The technical features are compatible with verbal dialog, and downloading applications and video in low definition.

2.5.1.3. *The Internet of Things*

The IoT is the extension of an Internet network for exchanging information and data from recreational or professional (industrial or medical) devices to the Internet network. The IoT affects low power equipment, and equipment with long operating lives or with limited functionality, where the bandwidth is weak and latency periods are tolerated (see section 2.7.1.2).

2.5.1.4. *Tactile Internet*

A new network defined by the ITU in 2014 as an Internet combining low latency, a very short transit time, high availability and great reliability with a high level of security, the Tactile Internet is aimed at the areas of industrial automation, transport systems, health systems, education systems and serious gaming systems. It has been designed for activities with a high level of sensitivity and precision across distances under 300 km in less than 1 ms. The quick reactions permitted by the Tactile Internet are the source of its name, evoking the tactile human sensitivity that bypasses the brain and is almost immediate.

2.5.1.5. *Darknet and the Deepweb*

The "Darkweb" is the worldwide Web system that includes Darknets and public Internet. Accessing the Darkweb requires specific software, configurations and permissions. The Darkweb also accommodates the Deepweb, which is not indexed by search engines.

A Darknet, sometimes known as a "Friend to Friend" network, is a small virtual private network established between around 10 trusted users. A "darknet" is often an illegal P2P network, using Tor (The Onion Router) and established for specific, undisclosed objectives. Connection times are long on the Darknet due to the existence of numerous changeovers. Initially, this phantom Internet, which guarantees anonymity, was created to help freedom of information in countries where everything is restricted, such as China. Dissidents can, via the Darknet, communicate in a secure manner. Anonymity and concealment can also easily be turned to dishonest ends: arms sales, pedophilia and drugs abound on this hidden Web. Tor is one of the most used pieces of software for surfing the Darknet. It is thought that there are 500 times as many Web sites on the encrypted Web as on the traditional Web.

The "Deepweb" is a part of the Web that is accessible online, but is not indexed by hyperlinks on ordinary search engines (except through the BASE browser). Some of these Web sites are not linked to other pages and they therefore cannot be discovered with indexing tools. It would be difficult to detect the wrongdoings carried out by some Internet users on the Deepweb. Accessing some Web sites can also be prohibited through the use of non-standardized document formats (opaque web). Other Deepweb sites have restricted access due to blocking organized by placing a file at the root of a Web site, banning responses to ordinary browsers.

The Tor network (The Onion Router) is made up of a group of servers that allow users to increase the level of Internet security through a series of virtual tunnels protecting their privacy. The Tor browser, making all Internet exchanges based on TCP protocols anonymous, is an effective circumvention tool, allowing users who develop software to create new communications tools that respect privacy. A sanctuary for political victims, Tor has also been the chosen refuge for criminals from the contraband and drugs markets. For this reason, the Memex search engine, capable of exploring the dark side of the Tor network, has recently been developed. Today, the Tor network is no longer only a hideout for criminals. In November 2014, the social network Facebook was set up there, probably for those users enthusiastic about anonymity, and, at the end of 2015, it was joined by the American investigative Web site “ProPublica”, created in 2007 by a non-profit organization, which is recognized in its field and holds two Pulitzer Prizes.

2.5.2. *Organization of the Internet*

The Internet is a global network accessible to the public. In 2015, the Internet network was made up of almost 70,000 independent networks, both public and private. Internet networks are interconnected through connections and routers. A message is relayed from router to router in successive leaps, with the router reading the message to be transmitted at the head of the packet and guiding it according to routing protocols. This packet switched data technology is carried out with the assistance of TCP/IP an UDP/IP protocols (layers 3 and 4 of the OSI model). Internet access is obtained through an ISP via various wired (ADSL), optical or radio technical means (see section 2.2.2).

ISPs offer flat-rate subscriptions to customers. After subscribing to an ISP and purchasing the necessary hardware, the customer accesses the Internet through their switched network telephone line (narrowband access or ADSL access with a faster connection). The Internet can also (see sections 2.2 and 2.3) be accessed through fiber-optics (FTTx or HFC), in CPL, or via radio link (satellite, 2G, 3G, 4G, WiMAX or Wi-Fi).

2.5.3. *Organizations involved*

2.5.3.1. *ISOC*

The Internet Society (or ISOC), created in 1992, distributes information and facilitates economic and social development by establishing reliable and lasting connections around the world through an open Internet accessible to all. The Internet Society ensures that neutrality rules are well respected by the different actors in the data transmission chain. It acts at all levels to ensure respect for rules, regulations and good practices. The Internet Society France

(ISOC France) is an organization responsible for managing the incoming and outgoing Internet connections of operators established in French territory.

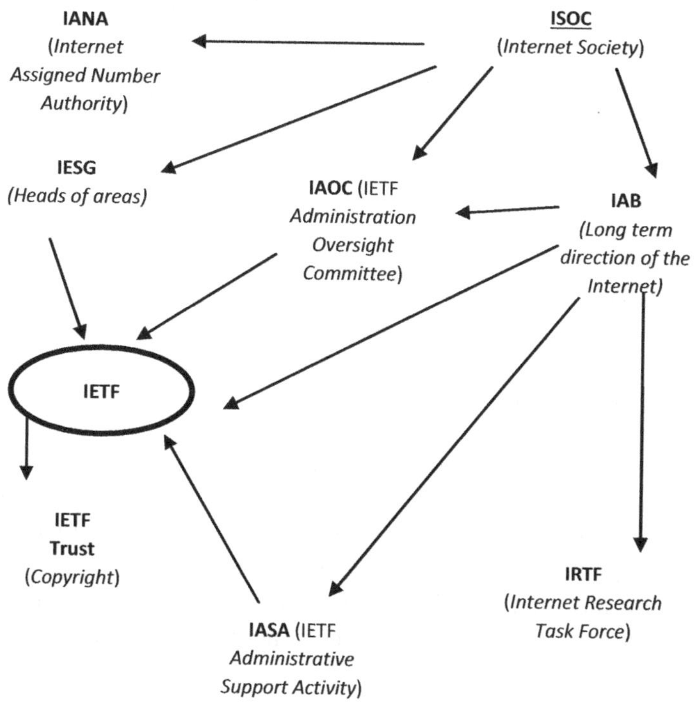

Figure 2.10. *IETF relationships*

2.5.3.2. *Internet engineering task force*

Created in 1986, the Internet Engineering Task Force (IETF) is a body that acts in the international legal field in association with the Internet Society (ISOC). Its objective is to provide Internet networks with communications standards and protocols, ensuring the perfect interoperability of equipment from different manufacturers. It gathers together a community of network specialists, operators and manufacturers, divided into over 70 working groups with interests in the architecture and changing operation of the Internet. IETF decisions are taken according to the "rough consensus" rule. Adoption is considered to be gained with 90 or 95% in favor for serious matters or merely 70–80% for less important topics.

The IETF area directors (AD) appoint the directors of their working groups, who belong to the Internet Engineering Steering Group (IESG). The areas studied mainly concern the applications, operation and management of the Internet network, real time applications and infrastructure, routing, security and transport.

2.5.3.3. *Internet architecture board and internet assigned numbers authority*

The Internet Architecture Board (IAB) is responsible for the long-term direction of the Internet and activities entrusted to IETF. The Internet Society (ISOC) governs IESG and IAB. All ADs and the president of IETF make up IESG, which is in charge of all management of IETF.

Internet Assigned Numbers Authority, a constituent part of Internet Corporation for Assigned Names and Numbers (ICANN), coordinates the assignment of values to IP parameters and distributes blocks of IP addresses to regional Internet registries (RIRs). The IRTF works on developing the Internet of the future, in coordination with the other working groups in charge of protocols, applications, architecture and technology.

2.5.3.4. *ICANN and NETmundial*

ICANN is a private American company, governed by Californian law and currently controlled by the United States Department of Commerce, which was created in 1998 to regulate the Internet. ICANN allocates root domain names (generic domain names such as .com, .fr, .org, in the addresses of Internet users and Websites) and keeps them in a dedicated database (the price of a domain name is around $200,000). In 2011, ICANN had a budget of $95 million, in connection with domain name registration fees. Naming changes IP addresses into domain names and vice versa. In the address <www.votre-entreprise.fr>, ".fr" is managed by ICANN.

"votre-entreprise.fr" leads to a server number managed in France by AFNIC.

A new ICANN, in the process of being organized, could resemble a government, with its own laws, purpose and moral values, and different categories of interest groups with certain powers, such as the ability to dissolve its administrative council or block its budget. The ongoing negotiations have presented some difficulties in defining the relationships between the new ICANN and State governments.

On the other hand, a parallel organization to ICANN has been proposed under the name of NETmundial. Members of ISOC, IAB and IETF want the structure of the NETmundial Initiative (NMI) to be defined, in addition to its mandate and area of jurisdiction. The coordination work must be continued between NMI and the different communities involved (https://www.netmundial.org/).

2.5.4. *Internet addressing and e-mail addresses*

On the Internet, the packets transmitted contain the address of the recipient's terminal. This IP packet addressing is sometimes carried out in relation to the IPv4 rule (version 4), which has become obsolete. The new version, IPv6, is being gradually implemented. To facilitate exchanges during the transition period, it is recommended that two types of addressing be used. Since 1992, ICANN has entrusted Internet addresses in Europe to the European RIR (or RIPE NCC, located in Amsterdam). They are the service providers who provide their customers with their IP address and e-mail addresses.

2.5.4.1. *IP addresses and domain names*

An "IP address" is an identification number allocated to every device connected to an IP computer network. The IP address is used for routing the packets transmitted across the Internet network. The IP addresses of terminals and servers can be converted into a domain name and vice versa. The domain name system (DNS) establishes the rules for writing these addresses in alphanumeric characters. The

Internet uses almost 300 top-level domains (for example ".org" and ".fr"), which are defined by ICANN.

The DNS translates the relevant domain name into the IP address associated with it. The URL, or "hyperlink", is a system for identifying the universal address, which allows a computer to connect to another computer on the network (URL example: http://www.sncf.com/). The DNS is a fundamental link in the architecture of the Internet. Considerable efforts have been made to secure it, such as the introduction of cryptography to confirm responses with DNSSEC.

2.5.4.2. *E-mail addresses*

An Internet e-mail address is made up of ASCII characters (in Asia, in Unicode characters). The addresses contains three parts: a personal part (alice.berteau), a delimiter (at, @), followed by the domain name of the root name server's host, and the top-level domain code (TLD), which may be generic, "gTLD", or geographic, "ccTLD" (country-code Top-Level Domain). The resulting address <alice.berteau@wanadoo.fr> is used by the mail server.

2.5.4.3. *IPv4 and IPv6 addresses*

Version 4 IP addresses use 32 bits (4 bytes) and version 6 ones contain 128 bits (16 bytes). IPv4 is no longer allocated owing to a saturation point in assignments having been reached and all Internet networks are invited to use IPv6. As IPv4 and IPv6 addresses are not compatible, either a dual stack operation or the implementation of basic mechanisms to translate and/or encapsulate protocols should be considered.

In the so-called dual stack method, network equipment (from level 3) must contain IPv6 (built-in) and IPv4 addresses to make unrestricted communication possible.

Translating protocols can take place at several levels of the TCP/IP stack (level 3, IP – level 4-transport – or higher levels, application).

Two types of solution are feasible: the tunneling connection (an IPv6 server communicating with IPv4 infrastructure) and the translation of IPv6 protocol to IPv4, which is associated with translating IPv4 protocol to IPv6.

None of the existing transition techniques are entirely satisfactory and integrated IPv6 is highly recommended. The translation of IPv4/IPv6 addresses with NAT equipment is advised against as it prohibits the installation of new applications. Over a long period of time, the transition between the two types of addressing will have to be taken into consideration. The method of transition depends on the Web sites and the configuration. The technical challenges of implementation are numerous in this case. Some applications require an immediate move to IPv6 (IoT for the medical and automotive industries, for example).

Implementing IPv6 addressing in business is expensive and sensitive. The shortage of IPv4 addresses has led to all equipment installed undergoing address alterations with regard to the links they are connected to, beginning as quickly as possible and under the best conditions. The best solution is to reproduce the working configuration in a trials network that the experts can easily alter. Among the main difficulties encountered, upgrading protection devices (firewalls), "multihoming" (simultaneous connection to several ISPs to improve the reliability of an Internet connection), adjusting the memory size of equipment to match IPv6 and adapting docking accessories to network discovery should be noted. The assistance of specialized IPv6 experts is highly recommended.

Transition costs will probably increase more and more over time due to the growth in the amount of devices to be taken into consideration. However, some experts believe that the resale of IPv4 addresses is possible and could be a palliative for small installations. Nonetheless, over the longer term, IPv4 will become obsolete in protocol stacks. Generally considered incidental, the cost of two successive interventions to change addresses has no economic benefit.

2.5.5. *Connections between IP networks*

The coordination of managing all connections used in IP is determined by the P2P, or peer-to-peer, connection agreements (so-called "peering" agreements).

2.5.5.1. *RIR and local Internet registry*

An RIR (of IP addresses) is a private sector organization under the authority of ICANN, whose members are local Internet registries (LIRs of IP addresses), network operators and ISPs. LIRs are contractually linked to their RIR and pay an annual subscription.

RIRs allocate Internet resources, namely IPv4 and IPv6 addresses, and autonomous system numbers used for routing traffic between operators with the border gateway protocol (BGP). They maintain the public routing database (Routing Registry) according to instructions received from LIRs. They distribute documentation and information related to addresses. The Internet is managed through five RIRs or regional coordination centers:

– RIPE-NCC (*Réseaux IP Européens* – European IP Networks) for Europe and the Middle East;

– APNIC (Asia Pacific Network Information Center) for Asia and the Pacific region;

– ARIN (American Registry for Internet Numbers) for North America;

– LACNIC (Latin America and Caribbean Network Information Center) for Latin America and the Caribbean;

– AfriNIC (African Network Information Center) for Africa.

In order to improve management of their network, every actor or participant is invited to analyze their network traffic. Internet switches (Internet eXchange Points [IXPs] or Global Internet eXchanges [GIXs]) are connected to each other (440,000 routes in 2013) so that service can be guaranteed in the event of broken connections. Exchanges of Internet traffic between networks are organized on the basis of mutual agreements, known as "peering", (equal to equal) which aim to reduce costs and latency periods and to optimize efficiency.

After consulting routing tables, routers transmit the traffic received from the nearest neighboring router to the destination according to the traffic information received. Extending the Internet often requires the creation of new IXP, which makes the routing tables and rules used by internal network protocols more complicated.

2.5.5.2. *Peering*

"Peering" is an operating mode that consists of exchanging Internet traffic with pairs, namely technically compatible pairs or those permitted through a commercial or previous contractual agreement. "Peering" also allocates a new type of virtual private network managed by an operator specializing in professional usage. It is a service connecting shared and superfluous platforms that allow the exchange of public or private traffic between authorized customers, or "peers", or between one "peer" and route servers.

A code of conduct determines the conditions to be respected by users (minimum traffic exchanged, private or public connections, traffic localization). The infrastructure is set up in one or several host centers or equipment and the proper user interchange infrastructure is installed. In order to guarantee a high availability of services, the infrastructure in place is duplicated so that the traffic is not affected by the risk of power outages.

2.5.5.3. *IXP/GIX*

Exchange points, both national and international, deal with public and private traffic in IPv4 and IPv6 for ISPs, content hosts and the other types of network that make up the Internet. These exchange points are the result of a community initiative from national operators, within the framework of an associative structure guaranteeing the neutrality, longevity and continuous improvement of the service. To connect two networks, transit procedures (paid for between meshed networks) or "peering" procedures (free or regulated via "peering points", known as "IXPs") are used.

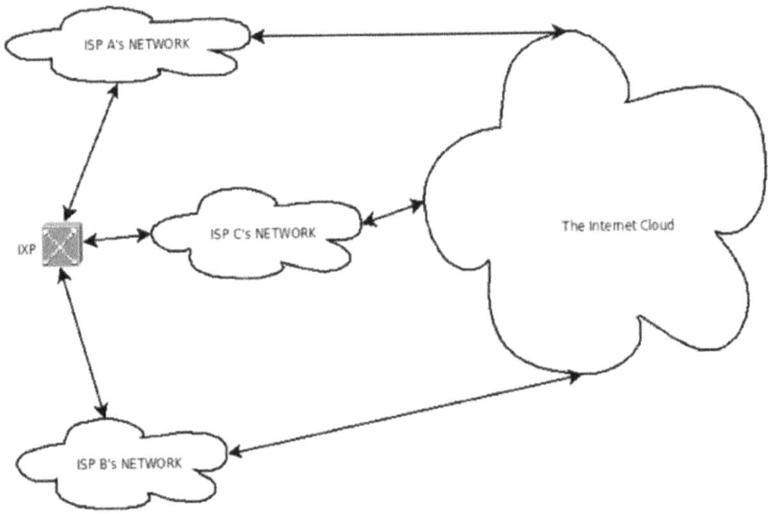

Figure 2.11. *Networks connected through IXP*

Despite the significance of their advantages, IXPs only exist in half of the world's countries and, when they are present, many do not even operate to their full potential. Indeed, their full roll-out is linked to the establishment of a favorable environment. An IXP is a unique physical network infrastructure operated by one organization, whose objective is to facilitate the exchange of Internet traffic. It serves as a centralized platform that allows local traffic to be routed and to conserve international bandwidth, which is reflected in a reduction in the global cost of Internet connectivity. The IXP is a regional exchange point that allows the costs of international Internet connectivity to be reduced. The IXP provides a connection platform for IP services between fixed-line and mobile networks and service providers. It guarantees greater security, reliability, scalability and monitoring. Through its guaranteed levels of service quality (SLA), the IXP determines interconnection for 4G/LTE mobile roaming services, voice signals in high definition (HD) and in Rich Communication Services (RCS), secure cloud services and the entire range of available IP services on fixed-line and mobile networks.

The GIX is an Internet exchange hub where different IP networks exchange their traffic via suitable equipment, thereby avoiding significant additional costs linked to data transport, while improving their bandwidth quality. These IXP/GIX points allow online services to be available to all users. They contribute to Internet development at local and regional levels (Figure 2.11).

2.5.5.4. *GIX example: LyonIX*

The Web site LyonIX is a neutral marketplace that allows the purchase and sale of Internet transit through direct connections between IP operators in the Rhône-Alpes region of France. LyonIX is an IPX/GIX open to non-professional Internet companies that require traffic resources more than 10 symmetric Mbps. LyonIX is a member of RIPE, ISOC-France and the *Fondation Internet Nouvelle Génération* (or FING – New Generation Internet Foundation), among others. LyonIX is a GIX-NAP:

– GIX, for exchanging IP packets or peering (the exchange can be made at level 2 – in BGP protocols – or at level 3 for route servers);

– Network access point (NAP) a connection point between different ISPs with peering for the IP protocols transit market.

2.5.6. *The concept of "Internet governance"*

The question naturally arises of who is responsible for the consistency of the Internet's technical standards, for maintaining all of its networks, and for planned extensions, as well as who is legally responsible for the form and content of services and applications. The topic of Internet "governance" has been discussed several times at international meetings without a clear solution being unanimously and definitively formulated.

Internet governance in 2016 seems still to be *de facto* carried out by the United States and ICANN, with the assistance of the organizations shown in Figure 2.10. It is intended that the circle of decision makers will soon be expanded to include universities and other bodies, with the United States not allowing other United Nations member States to join, or even ITU or UNESCO. The objective would be to implement stable and manageable governance while pursuing the initial global progress objectives of peace and the free flow of ideas and knowledge.

On a practical level, users are therefore defenseless as soon as the quality of service connections is called into question or intrusions into personal data are observed, as there is no effective authority of legal recourse. The operators, for their part, participate in collecting customers, traffic flows and network maintenance, without being able to affect the form and content of applications. There is a great risk of countries with “BIS Internet” opening to global traffic before too long (see section 5.1.1).

2.6. Digital network terminals

2.6.1. *Main models used in 2016*

There are three discernable categories among the most common uses: desktop computers or laptops, mobile devices with 3 or 4G technology (smartphones) and tablets.

2.6.1.1. *Desktop computers*

A desktop computer is connected to an external keyboard and a printer. Its screen is large and commands are made with a keyboard connected to a mouse. The computer is both a stationary working instrument and a terminal focused on data exchange. It is also possible to combine applications enabling line-to-line voice services with images.

2.6.1.2. *Smartphones*

A smartphone is both telephone and computer, and acts as a personal assistant, as well as being able to take digital photographs and send them across the 3G or 4G network to which it is connected. It consists of a thin plate with alphanumeric keyboard and a small touch screen.

Type of terminal	*Keyboard*	*Screen*	*Mouse*	*Voice*	*Multimedia data*
Computer	Yes	Large	Yes	No	Yes
Smartphone	12 Keys	Small	No	Yes	Simple data
Tablet	Touch screen	Medium	Stylus	Yes	Yes

Table 2.3. *Comparison of the types of terminal available in 2015*

Smartphones are designed for voice telephony, accessing e-mail, looking up information online and online shopping. They have considerable processing capabilities that, added to the Internet connection options, increase their power.

2.6.1.3. *Tablets*

Tablets have an increased number of smartphone functions and manufacturers are in search of the ideal model, combining a small, sophisticated smartphone with two-way video and a tablet (known as a "phablet"). In principle, tablet use should be restricted to Internet data applications, and should not allow telephone conversation. However, some industrial developments unveiled in 2016 have shown that this limitation has been overcome.

2.6.1.4. *Mobile telephony*

Be it a smartphone or a tablet, mobile telephones are an indisputable factor in economic development. They allow voice or written communication, money transfers, even in the most disadvantaged geographic areas, and access to information, including aspects related to personal health. Secure online payment applications encourage innovation on smartphones and tablets.

2.6.2. *Selling or renting a terminal*

When the communication terminal is provided by the network operator, the operator is authorized to ask the customer to pay a rental fee every month to be added to usage costs. In such cases, the operator agrees to take on repair or replacement costs. Most of the time, the terminals offered for customer rental are reliable and the customer can generally pay up to 2.5 times the terminal price during its operating life. In return, the customer benefits, without additional charges, from the advantages presented by the automatic updating of service offers compatible with their terminal, and even its annual replacement.

A simple mobile terminal, rented for €7 per month, amounts to €420 after 5 years, whereas a 4G terminal with greater functionality is sold commercially for €440. Users can exercise their choice depending on their options and offers that today are ever more appealing.

2.6.3. *Bring your own device (BYOD)*

In a professional context, this practice poses problems in acquiring and operating a terminal.

In acquiring a personal terminal, employees wait for their employer to provide them with suitable technical support and financial compensation for professional uses. The company is not always interested in the initiative taken by its employees. Indeed, the BYOD practice involves having several versions of different terminals in the same company that presents various security risks. For their part, terminal providers have one contract (provision and maintenance) from the time of subscription.

In company operations, computer systems administrators prefer to remain in charge of the inventory of mobile equipment (computers, telephones, tablets) connected to the company's IT service, as well as the management software of mobile devices. The BYOD practice introduces confusion to the character and encryption of the data carried by the terminal (personal or company data). The use of BYOD heightens the risk of attacks, whether by intrusion, or by passive surveillance of transmitted information. In some professional areas (the medical sector, for example), it is impossible to prevent the use of personal telephone or computer terminals in the workplace. The BYOD concept remains a challenge.

In 2015, some companies, including the Orange Group, decided to provide all of their employees with a professional mobile terminal.

2.7. Internet applications

This section summarizes the characteristics of the principal network applications identified in 2016. The diversity of their commercial targets is combined with the technical characteristics linked to the average volume (in bytes) of data exchanged, the requested level of security, and the optimal transit time that makes it difficult to come up with a strict pricing system on the basis of the provided elements of service. Network operators often use a monthly rate base that does not always relate to the service provided. The majority of application operating revenue is not taken by network operators, but by application originators.

2.7.1. *The Internet and the Web*

The Internet and the Web use different technologies and have different functions. However, these terms are often used synonymously.

2.7.1.1. *Internet*

The Internet today is a network made up of around 70,000 public, private, university, commercial and government networks. It is a "global network connecting telecommunications networks to computers, in order to allow the exchange of electronic messages, multimedia information and files". The Internet is a system shared geographically and structurally for publishing and consulting documents using hypertext technology (computer cross-referencing system) (Official Journal definition on March 16, 1999) [KIR 11].

2.7.1.2. *The Web*

The Web (World Wide Web) is the Internet's principal application. It is a system for consulting multimedia documents (texts, sounds, images), which uses cross-referencing to allow one part of a document to link to another, or to connect a document to others considered relevant.

The Web implements client–server operation. The Web offers a directory service, the option to conduct targeted searches of all Web sites and to create a web journal (known as a "blog"), as well as participating in forums, online games, online shopping (e-commerce), and exchanging electronic mail (webmail), among other things. From its installation on a server, a Web application is immediately usable by any user in the entire world who has a computer connected to the network and is supplied with suitable client–server software.

The World Wide Web allows users to share documents and images more quickly than Internet, as it does not use client–server communication, but P2P exchanges, which are faster (see sections 2.7.2 and 4.4). The Web also makes typical applications, such as the "Rich Internet Application", more manageable. Generally speaking, it must be remembered that an application server is a computing device connected to the Internet and used for hosting simple or multimedia applications.

2.7.2. *Mode of operation*

2.7.2.1. Client–server

For over 30 years, the protocols applied to the Internet have enabled highly used applications, such as data transfer, e-mail and instant messaging, which use the principle of client–server exchanges on the basis of specialized protocols. The "client–server" mode of operation is used between the user terminal and the network server on the TCP/IP protocol stack. Alternatively, the user sends requests to which the server responds. The mode of operation is implemented for connecting to a browser to search Web sites and display pages (in HTML or XML Socket), file transfers (with FTP) and messaging (with SMTP, POP and IMAP).

2.7.2.2. *Peer-to-Peer*

"Peer-to-peer" communication handles the exchange of multimedia data based on different client–server principles, with the computer and Internet user requiring specialized software capable of establishing communications with callers using only its nearest network hubs, without passing through a central server. Every computer is both client and server (hence the exchange name being from one equal to another, P2P). The best known software programs are without doubt BitTorrent, Edonkey, Emule, Kazaa, LimeWire, Napster, Skype, Spotify and TeamViewer. P2P applications allow file sharing, distributed computing, IP telephony and sharing multimedia feeds in streaming (see also section 4.4).

2.7.3. *Voice applications and voice messaging*

Voice signals transmitted through the PSTN must be differentiated from those transmitted digitally, namely in HD VoIP and from those in (non-standardized) VoIP. VoIP-HD is often offered on mobile telephones.

The PSTN allows voice exchanges in the low range of frequencies transmitted and the signaling system set up through the terminal keyboard provides the option to choose services on the recipient's reception system, for example leaving a voice message on an answering machine that will be accessed later. According to ARCEP [ARC 14] statistics, the monthly volume of PSTN calls fell 8% in 1 year, reaching 3 h 58 min in the first quarter of 2015, as opposed to 4 h 19 min a year earlier and 5 h 46 min in the first 3 months of 2012.

Voice messaging was still fairly rare in Europe in 2015, while it has become commonly used in Asia and South America. The social network WeChat has 600 million members and WhatsApp subscribers seem to be interested in this mode of personalized exchanges, which do not require spelling knowledge.

2.7.4. *Text messaging*

E-mail was created in 1971 on the Arpanet network and was popularized by Compuserve in 1992. Initially, each element of the message was charged, including retransmission to a third party. In 2016, the price of the messaging service is included in the Internet connection package and this free appearance explains the growth of personal and professional messages (200 billion messages per day). Messages between parties can be in text mode, as the alphabets used are identical, either through messaging systems (short message service (SMS) or e-mail) or through a choice of social networks. Instant messaging applications allow a dialog to be opened with consumers (Table 2.4).

	Free SMS messages	*Coverage*	*Number (millions)*	*VoIP*	*Video*	*P2P*	*Funding*
WhatsApp	*	Am, As, EMEA	1,000	*		*	Subscriptions
Facebook Mess	*	Am, As, EMEA	800	*	*		
Viber	*	As, EMEA	250	*			
Skype	*	Am, Europe	350	*	*		Business
Yik Yak	*	Am.	?	*			3 km GPS
Kik	*	Am, Saudi Arabia	100				Ads
Snapchat	*	Am.	200		*		Ads
Telegram	Encrypted	Am, As, EMEA	100	*			Donations
Line	*	Am, As	170	*	*		Games
WeChat	*	As, Africa	470	*	*		Store
Kakao Talk	Mobile *	Asia	35	*	*		Games
Tango	*	China	250	*	*		Ads, games

*Yes.

Table 2.4. *Different instant messaging applications [Source: NEXMO]*

WhatsApp, acquired by Facebook, with a billion active users per month, is the leader in Europe and the United States. Applications such as WeChat, Line, Kakao Talk, Viber and WhatsApp currently have over a billion active users per month between them and rank among the ten most used applications in the world, according to Mary Meeker of Kleiner Perkins Caufield Byers. WeChat dominates in

China. The Line messaging service is growing in Japan and several south-east Asian countries, Kakao dominates in South Korea and Viber is active in various countries. In Asia, instant messaging platforms, such as Line, WeChat or Kakao Talk, offer business, online payment and information services or a range of video games. These messaging services wish to extend their influence in order to position themselves as central platforms through advertising, covering all of daily life, including horoscopes, weather forecasts and taxi bookings.

2.7.5. *Searching for information online*

The protocol-based client–server system favors searching for information on databases, which are semipermanently available on the Internet worldwide. In this case, it is likewise only possible to exchange information if the typographical characters are the same. In this question and answer service, all topics can be addressed in return for some estimation and adjustments made to texts without prior notice. The process brings to life the saying "I click and I learn everything", and the "copy and paste" practice, opening the way to the loss of author rights and Copyright. The Wiki Web sites (from the Hawaiian word for "quick; fast" and that can also be translated as "What I know is") have software on their servers, which authorizes visitors to edit the pages they visit, thereby paving the way for "participatory management". Alongside collaborative Wiki Web sites such as Wikimedia commons, Wiktionary and Wikibooks, the Web site Wikipedia has been, since its creation in 2001, one of the biggest, unofficial online encyclopedias and is usable in almost 280 languages (500 million unique visitors were recorded in January 2016), thanks to crowdfunding and the support of Google and Microsoft.

Remote, personalized medical advice has existed for several years and is currently regulated. However, the Internet presents issues surrounding the reliability of information and this activity can contribute to spreading anxiety. Doctors, contacted by telephone and e-mail, are faced with this competitive practice. Hypochondria is a psychiatric condition, known since the times of Hippocrates and Molière, which is characterized by the belief that one has contracted a fatal illness. INSEE estimates that "cyberchondriac" activity has increased from 26–41% among the French population and could be linked to the growth of Internet use, as it is ranked as the third most common activity among Internet users, with a predominance of women and young people. In France, one in three Internet users has accessed health information from a mobile platform (one in five in the United States).

Let us recall that the HON Code, credited to the Swiss foundation of the same name, (Health On the Net Foundation, <https://www.healthonnet.org/ >) is a charter with the goal of guaranteeing that health-related Web sites respect the principles of good moral conduct, but it does not include the complete truthfulness of information put online (section 4.2.3).

2.7.6. *Videos*

2.7.6.1. *Audio-video messaging*

For users of audio-video messaging applications, the quality of communications is essential. Furthermore, Google uses P2P connections to improve performance where these connections are available. The risk can be doubled as P2P slows down operations and allows the IP addresses of callers to be disclosed.

2.7.6.2. *Streaming*

The continuous transmission of images (streaming) with the appropriate sound is possible on NGN in low and medium definition. Video transmission protocols are available and allow live or delayed exchanges for private communications or those with a commercial or artistic purpose. Social networks use these technologies and can be used to spread general or specialized training courses (massive open online course, MOOC).

2.7.6.3. *Television*

The creators of television programs want to stay in touch with their viewers by using the resources of hand-held devices and the availability of high-speed mobile connections. The offer of video on demand (VoD) has recently been followed by packages of television shows being broadcast on computers or tablets, either by subscription or for free, particularly on the occasion of major sporting events. The addition of strong advertising is a path currently being explored for this application, often called "connected TV".

This part of the market is being reorganized. On the one hand, the frequencies available for DTT television broadcasts are few and far between and, on the other hand, commercial breaks are more expensive there than on the Internet. Furthermore, the BBC distributes programs on the Web, which is preferred by young people in the early evening. Not all countries are affected by such reassessment of carriers.

2.7.6.4. *Streaming television*

The Internet enables television shows to be broadcast through streaming (continuous flow of images that are stored for several seconds before appearing on screen in order to disguise the slight delays that can occur during transmission). The regulatory authorities have nonetheless requested that network operators do not block or slow down these online video services for reasons relating to technology or competition ("Net Neutrality", see section 4.2.5).

Netflix, the American subscription VoD service, is interested in the French- and German-speaking markets for broadcasting films and series on the Internet. It has an extensive catalog of TV series and also produces original programming. These programs can be accessed through high-speed channels on adaptors for connected televisions, as well as on tablets, smartphones, games consoles and computers.

2.7.7. *Online music*

Over the last few years, there has been a drop in the average expenditure in the music sector, linked to the growth of the market. According to the consultancy firm Bain & Company [BAI 15], in western markets, a user's monthly spend, which was $8 when the CD was the market leader, has plummeted to $3 with downloads. With streaming services such as those of Spotify, it is no more than $2, although it is hoped that there will be a reversal of this trend.

2.7.8. *Personal data*

Content producers have taken on the practice and the right to host and use the personal data of Internet users who visit their Web sites. Between one-third and one-fourth of Internet users are willing to share their personal data with Internet providers. Reluctance is all the greater given the fear of a harmful use of this information. If seeking to monetize this practice and guarantee data revenue with the help of a computer system as with Big Data, it would be necessary to convince customers that their data were protected and would only be used to assist them. The desired economic model works on the assumption that any security-related incident would not cause the model to come crashing down immediately.

2.7.9. *Online purchases*

Online shopping (also called “e-commerce” or “distance selling”) is the result of facilities offered by the fast interactivity of connections bringing together a potential buyer and the database of a remote seller on a digital network. Electronic data exchange enables choice and allows buyer identification. The system that has been implemented includes a remote payment stage, connected to electronic banking devices. Its ease of use and efficiency mean that e-commerce is currently primarily mobile-based, indicated by the term “m-commerce”.

2.7.10. *Geolocation*

Geolocation (or providing a reference position, GPS) is an application that allows a device or person to be located through geographic coordinates. There are several means available that can sometimes be used to provide greater precision, such as geolocation through Internet IP address, geocoder, satellite, GSM or RFID radio-based communications systems or Wi-Fi. Remote reading of information is another application linked to geolocation information.

2.7.11. *Social networks*

Social networks are communications services that bring together groups of Internet users on the basis of social or professional connections, or ties of friendship. The content exchanged between users has been the object of sociological and commercial studies whose conclusions have been used by the managers of these social networks. It has thereby gradually developed that after the free phase guaranteed through advertising and the highly profitable resale of personal data, these managers have become interested in the commercial potential of online video services, remote payment services and remote shopping services. Among these, Amazon, Instagram, Linkedln, Myspace, Twitter, Viadeo and YouTube, among others, must be mentioned.

As an example, Facebook, with over 1.5 billion users, has become a global service. "It offers a large range of personalized information services, ranging from virtual reality for the medical sector to online purchases with online payments, and from university education to recreation. It is more open to marketing than Google or a TV channel due to its progressive opening to customers, which has attracted 24 of the 33 million French Internet users and almost half of mobile Internet users, whose personal data are collected in vast semantics database without necessarily being used" [KIR 11].

In controlling an important part of culture and the economy without referring to network operators, these new OTT Internet players can help to challenge the standard practices of national institutions.

2.7.12. *The Internet of Things*

The IoT is seen as the third phase of the Internet (or Web 3.0, or Semantic Web), after the Social Web (Web 2.0). It is the logical evolution of "Big Data" and "Cloud Computing" applications. The IoT uses IPv6 addressing, taking into account the significant number of things to be connected and a specialized network with a slow transit time. It ensures information exchanges coming from physical devices to information collecting systems and remotely controlled systems connected to the Internet (see section 2.5.1.3).

The potential IoT areas are e-Health, robotics (RFID, M2M), home automation, Smart Cities, an Intelligent Transport System, energy management (Smart Grid) and personalized well-being assistance (Quantified Self), as well as industrial areas. All of these developments could require the connection of over 50 billion things between now and 2020 and could considerably alter city management and the lives of citizens.

Several competing IoT connection systems are currently being developed. SIGFOX is being rolled out in the United States within the framework of smart city programs and industrial applications, and Arqiva is building a national IoT network with SIGFOX technology in the United Kingdom, France, the Netherlands and Spain. The low power radio technology LoRa, a competitor of Sigfox, bridges the gap between short-range radio technology (Wi-Fi, Bluetooth, ZigBee) and the longer range technology of mobile networks.

2.7.13. *Banking applications*

The most used mobile telephone applications are those related to banking. Banks are focused on the most common uses, such as consulting account balances, transfers from one account to another or external transfers, and identifying customers from the moment they begin using the application. The development of new functions is planned to occur through SMS, such as the notification of overexpenditure or bank card fraud by SMS, transferring a customer's spending predictions, and the possibility of working directly with an advisor using the mobile application (see Appendix 11 on MFS, mobile financial services).

2.7.14. *Online gaming*

An online game is a mechanism that connects players (two or several) on a computer network, such as the Internet, through software allowing them to enjoy musical broadcasts and videos. Online games can be accessed for free or for a paid subscription. Marketing and advertising can provide financial support for the management of this activity. The latter is particularly interested in Internet users between the ages of 18–35 and concentrates on around 400 successful games, which represented a total in the range of $7 billion annually in 2013. Online games are played on different types of device ("playstation", central consoles, computers, tablets with large or small screens). Numerous online games can bring about the creation of virtual communities, opening up the possibility of enlivening social activities. Since 2010, the hosting of paid games and online gambling activities in France has been supervised by strict regulations.

Gaming platforms allow us to observe how players react and how they organize themselves into action groups, mirroring what happens in businesses. The roll-out of virtual games allows players to be observed psychologically and sociologically, regarding their taste for risk, compliance and non-compliance with rules, and so on. Multiplayer games are moving ever closer to the real world economy and have their own financial crashes, to the extent that editors employ economists to manage the thousands of daily commercial transactions generated in their video games (see www.engagdget.com/ and www.inaglobal.fr/).

2.7.15. *The Cloud*

Cloud computing is a computing application that connects computational power to demand via through the Internet. Its flexibility allows a large amount of data to be processed, whatever it may be, in a relatively short time at a relatively low price. This outsourcing of services commits the company to long relationships with its contractors. The security of data entrusted to the manager and the business's own vulnerability to attacks leads to data being duplicated or to working with two partners to reduce vulnerability. Cloud computing technology is increasingly used in commercial enterprises.

2.8. Overview of network connections

2.8.1. *Distribution network*

The collection of professional and residential customers is carried out according to the principles stated in sections 1.4 and 2.2.2. This collection must be understood as the sum of the actions of different participants, be they competitors or associates, who work in the same field. For each of them, the staggered repayment of investments made takes into account all of the work and equipment involved in servicing fixed and mobile technologies, which may or may not be used.

2.8.2. *National network*

At the national level, the network consists of a mesh network where each network hub and each branch is able to partially relieve the traffic flow of hubs and nearby branches (in principle, up to 30%). The most important hubs connect the fixed-line network to mobile communications networks and to the Internet.

2.8.3. *International connections*

Every country is connected to its neighbor by two types of connection, a bundle of terrestrial links (sometimes in addition to high-speed radio links) and a bundle of high-speed or very high speed Internet links (see Figure 2.8). "Land" exchanges are subject to negotiation and regulations between partner countries and Internet connections are arranged between the beneficiary country and the national or regional branch of the Internet Society (ISOC France, for example).

In the two situations, recourse is made to the international agreements signed previously and according to the rules defined by

ITU-T Study Group 3, as mentioned in sections 3.4–3.6. The parameters used on this occasion are related to the bilateral flows observed in the connections and the equipment set up on both sides.

2.8.4. *High and very high speeds*

In 2016, the regulatory authorities characterized the high-speed servicing of subscribers as where downstream access is under 30 Mbps. The very high speed relates to connections where the downstream capacity is over 30 Mbps. The boundary between low and high speed is fixed at 512 kbps. The government authorities' declared objective on the Digital Agenda is to provide coverage of 100% of French territory with very high speeds by 2022.

2.9. Network management staff

A network cannot be correctly operated unless its specially trained staff is continually involved from its creation. The skills of this staff must correspond to the network's features. Consequently, each significant geographic site must have experts in transmission equipment, energy workshops, signaling and network computing. The quantity and availability of these experts is dependent on the importance of each of the relevant Web sites and the rate of faults among the equipment involved. The expertise of these experts must also match the generation of materials used in the center of the network. The fast replacement of generations of hardware currently poses the problem of continually having to adjust the knowledge of staff on every site in the network. The examples provided below illustrate some of the difficulties that are likely to be encountered.

In the United States in the 1970s, network operators thought that frame relay (FR) technology was more appropriate and less expensive than protocol X.25 technology for routing traffic consisting of financial and stock exchange messages. To this end, they built private networks throughout North America. These companies prospered and multiple FR networks extended between the major stock exchanges until successive mergers between the managing companies were considered timely. Between 1985 and 1990, one final merger led to the unification of the last FR networks, justifying cutbacks to the staff of the operation and maintenance teams. This concentration of resources, rapidly carried out and linked to a far from complete transfer of knowledge, resulted in a gigantic failure that affected North American stock exchanges for several days. The new managers tackled the network relaunch and then set about reviewing the standardization of FR in order to remove the technical options likely to cause operational confusion. This anecdote shows the synergy required between operation and maintenance tasks and writing the standards for network equipment.

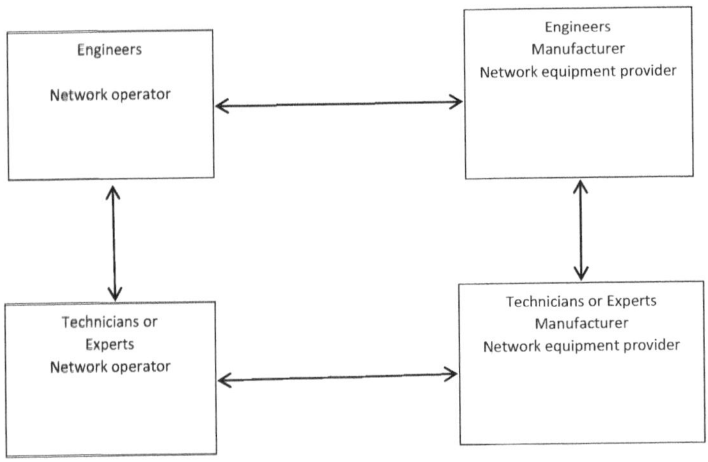

Figure 2.12. *Coordination between the four categories of technical expertise*

A similar example could be provided on this matter regarding helicopter maintenance. The proper functioning of a fleet of helicopters can only be achieved with the participation of specially trained expert mechanics equipped with suitable tools for the maintenance and periodic servicing of engines after every 25 or 50 h of flight. Similarly, Microsoft and IBM periodically "certify" their experts for very precise tasks adapted to suit clearly defined computers. Complementary training courses are periodically organized to take experiences acquired into account.

Certifying network maintenance agents, within today's highly unstable framework, poses many questions for managers. From 1970 to 1980, some countries in the Arabian Gulf managed their network through the intermediary of a subcontracting company, which recruited experts trained by the experienced British company Cable & Wireless. With communications networks using analog technology and their management being relatively uncomplicated, experts of all nationalities could carry out the necessary tasks without much difficulty. The idea of making the training of international network experts commonplace has become close to what the company Northrop established to respond to the demands of American military aviation. The multitude of standards and the appearance of digital technology have made this project impracticable. On the other hand, a communications network must guarantee the secrecy and confidentiality of messages, which has led to the recommendation that agents of the same nationality be used over long periods. Recourse to the civil service, generally undertaken in Europe before the 1980s, is the cause of these considerations.

In principle, these occupations are separate and relations between equipment providers and network operators remain distant. Similarly, in every profession, these jobs are structured in more or less strict hierarchies according to national cultures. However, it often the case that, in work involving international collaboration (regional projects or standardization work), the understanding of physical events was different among each of the four technical responsibility categories.

Experience has repeatedly shown that direct information exchanges, such as those shown in the diagram in Figure 2.12, produce very positive results in the studies conducted.

In a network's operating budget, the position reserved for staff is a significant portion that is difficult to make smaller without reducing levels of service quality, damaging continuity in correct functioning and guaranteeing the secrecy of information transmitted. Rapid technical changes have brought about a drastic reduction in the number of agents needed by a network, which was not easily foreseen. In some countries, the rules in force have allowed agents to be relieved of their civil servant status overnight. In others, the surplus staff have been transferred to other services without incurring excessive costs. In all countries, the career change requires training appropriate to new tasks, which is not always feasible. If, within networks, implementation tasks are being simplified, the organization of works requires good practical and theoretical knowledge in numerous fields, including computer science and security systems. The current division between network functions, operation of applications and the capabilities of devices poses a problem of how to globally organize technical user support. The spread of Internet access for all can only be made possible if a good level of quality of service is achieved, and if the ergonomics is genuine. Coordination must be guaranteed between these three occupations.

2.10. Assessments considered by network operators

2.10.1. *Internet revenues*

The short description of network structures given in this chapter shows that network operators draw their revenues from:

– subscriptions taken out with them by Internet users;

– telephone and communications subscriptions taken out by their fixed-line network customers corresponding to the relevant duration and distances;

– flat rate subscriptions taken out with their mobile network;

– the positive or negative balance of "peering" trade resulting from Internet traffic with partner operators and ISOC;

– and a percentage of the fees for subscriptions taken out on the Internet (via ICANN), with a deduction made for the purchase of IPv6 addresses.

It should be noted that the majority of activities described in section 2.7 (Applications) are currently offered to Internet users for free, with their managers finding remuneration in the advertising available online and the resale of the personal data files of Internet users to third parties. This situation should develop due to the possibility of implementing applications, with or without low definition video, connecting social networks to advertising, online sales and secure payments.

2.10.2. *Controlling content*

While a telephone network operator can block the nuisance calls it receives, this is not always the same on the Internet as an application can be managed by a service provider located in a country outside the network of the Internet user who falls victim.

The network operator who provides Internet access is unable to assist its customers in selecting the best Web sites and protecting them against potential attacks. At most, it can stipulate some general rules. The network operator is not able to control the content of Web sites, which its Internet user customers could access. This situation is especially delicate with regard to young Internet users. According to states and the general public, this lack of content control could be seen as a major freedom of expression or as a provocation with regard to national culture or local traditions. If some moderation of freedom of expression has seemed necessary on the Internet, its implementation is yet to be defined.

2.10.3. *Structural weaknesses*

NGN ensures the Internet's compatibility with fixed-line and mobile digital networks. Looking ahead, NGN allows access to all available applications on the Internet through client–server operation or through P2P. The Internet offers a vast range of possibilities for exchanging knowledge and information of all kinds. The ergonomics of Internet use and its quality of service requires the coordination of all actors involved (network, application, device).

In January 2016, the failure rate of connections to unicast IPv6 was nine times higher than the equivalent figure for IPv4 (1.8% of all attempts), but improvements are expected. On the other hand, the connection failure rate remains high because universal interconnection is not a requirement imposed by either the regulatory authorities or any voluntary arrangement between network operators. Connections between Internet networks do not obey the principle of network hierarchization, but the principle of free connections through "peering". "Hierarchization" refers to networks being tacitly structured into a hierarchized collection of links with customers and providers. The Internet is therefore not entirely interconnected and an asymmetrical gray area exists around its edges. Some areas are better connected than others, where there is a self-sustaining routing monopoly at their center. Other areas remain in which the packets sent never receive a response (http://www.potaroo.net/).

The absence of governance and a structured legal framework make the Internet an environment which is suited to attacks of all kinds and an ideal setting for networks of suspicious activity (mafia, subversive groups and criminal economy). These weaknesses will probably lead to the introduction of stricter technical and regulatory adjustments.

3

Communications Services Regulations

The goal of the international agreements concluded between states and bodies under the auspices of the United Nations (UN) is to ensure world peace by easing trade restrictions on goods and services and by removing barriers to international trade. Communications services must both satisfy the demands placed on services by financial regulatory bodies and agreed regionally and internationally between States, while also respecting the technical standards and rules determined by recognized standard setting entities. Large international banks take action in accordance with the trade agreements concluded between signatory countries.

Figure 3.1 summarizes the existing relationships between the technical organizations involved in the telecommunications sector and the different bodies connected to international political and financial institutions.

3.1. The international regulatory framework

3.1.1. *Trade agreements between States*

Since 1945, international trade has been governed by GATT and the World Trade Organization (WTO). In the absence of a unified world government, the United Nations Conference on Trade and Development deals with a trade system related to the development of countries in the Global South. Several economic agreements relating to regional economic areas have been arranged. Monetary policy is boosted by large international and regional banks through concerted efforts, such as variations in interest rates and the money available or even through questionable, unconventional policies (quantitative easing, changes to exchange rates).

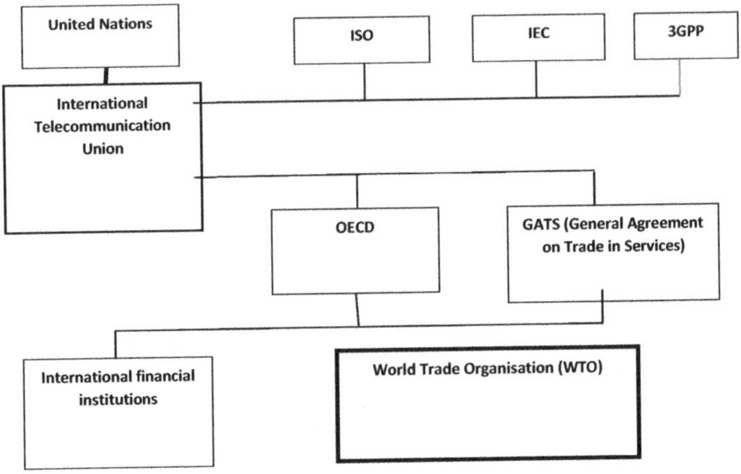

Figure 3.1. *Relationships between international bodies*

3.1.1.1. *The UN*

The UN seeks to establish world peace and it brings together the majority of States to this end. Its objectives are to facilitate cooperation, security, economic development and social progress. The UN Security Council is made up of five permanent members with the right of veto (the United States, Russia, China, France and the United Kingdom) and 10 members with 2-year mandates. The UN manages several specialized agencies to carry out its missions successfully, including the World Health Organization (WHO) for health, UNESCO for developing education, science and culture, the ITU for telecommunications standards, the Economic and Social Council (ECOSOC), and the World Bank Group.

The UN has not escaped criticism in its 70-year history. While it has saved lives and contributed to improving health and education around the world, its bureaucracy is excessive, undemocratic and too expensive. The enmity between poor and rich countries is glaring in the organization's lack of adequate representation, in the impact of decisions taken and in the outcomes achieved. The governance of the UN is considered increasingly ill-suited to the difficulties bombarding the modern world.

3.1.1.2. *The IMF*

The International Monetary Fund (IMF) is an international institution created in 1945, bringing together 188 countries with the aim of "promoting international monetary cooperation, ensuring financial stability and facilitating international trade". IMF funds are determined by quota subscriptions calculated according to each country's relative size within the global economy. At the same time, these quotas determine the maximum contributions to the fund, the weighting of votes within the Financial Committee and access to its finances. The qualified majority of votes are fixed at 85%, with the United States having *de facto* right of veto with 17.68% of votes. The IMF collaborates with the International Bank for Reconstruction and Development (IBRD) and General Agreement on Trade in Services (GATS). In 2015, it readjusted the currency composition within special drawing rights (SDRs) by introducing the Chinese currency (see section 4.3.1.2). IMF actions can at times compete with or complement the activities of international banks.

3.1.1.3. *The Organisation for Economic Co-operation and Development*

Since its creation in 1961, the Organisation for Economic Co-operation and Development (OECD), an international organization producing economic reports, and with a consulting role, has sought to promote policies aimed at achieving the greatest possible expansion in the economy and employment. Its activity aims to improve the standard of living in member States, while maintaining financial stability and contributing to the development of the global economy. Every year it publishes the main economic indicators of member States, including their gross domestic product (GDP). As of 2016, it had assembled 34 member States with democratic systems of government and which follow the tenets of market economics. OECD actions have not escaped criticism. The organization has been rebuked in particular for participating in actively promoting economic liberalism and global capitalism.

3.1.1.4. *The WTO*

The WTO, created in 1995, is an international organization, independent of the UN, that aims to promote international trade between countries. It brings together the 162 countries that, as of 30 November 2015, have signed GATS. Its main objective is to encourage openness in global trade. Its strategy has led to the negotiation of free trade agreements and the proposal of solutions to potential conflicts. China joined the WTO in December 2001.

3.1.1.5. *The GATS*

The GATS, signed by 153 countries and entering into force in January 1995, is the result of work carried out by the WTO and addresses trade liberalization and the fight against protectionism. According to some, GATS should increase the effectiveness of the means of production, as the growth of trade favors the comparative advantage of the countries involved. According to other sources, its actions could bring about the gradual destruction of public services.

3.1.1.6. *The G20*

The Group of twenty (G20) brings together 19 countries and the European Union. Created in 1999, after the financial crashes of the 1990s, the G20 aims to encourage international cooperation by introducing the idea of a broader dialog that takes into account the growing economic influence of several countries. It accounts for 85% of global trade, two thirds of the world's population and over 90% of the gross world product. The G20 supplements the economic cooperation actions mentioned in section 3.1.1 through regular meetings organized by the heads of State of member countries, the heads of central banks and other heads of State.

3.1.1.7. *The Davos forum*

Proposed in 1971 by Klaus M. Schwab, a professor of economics, the World Economic Forum in Davos has today become a non-profit organization, acting as an informal meeting platform for the main officials of public and private bodies who contribute to the globalization of the economy. Within this circle, the most pressing global issues, including health, culture and the environment, are discussed every year in a non-binding manner. The Forum also publishes a certain number of economic reports. It has acquired observer status at the United Nations Economic and Social Council.

3.1.2. *International financial bodies*

Trading relationships between States are the product of actions uniting several international or regional banks, agreements signed between countries and the actions of organizations specializing in international trade. The influence of these bodies can affect the entire world or regional groups of countries. The principle underpinning current international trade requires that the greatest possible degree of specialization in production activities simultaneously leads to a significant increase in trade, encourages competition, improves the quality of products, lowers sales prices and encourages technological transfers, all for the well-being and benefit of consumers.

The World Bank, created, like the IMF, in 1944, brings several international financial institutions together under the auspices of the UN, including the IBRD and the International Development Association, which were created to assist, advise and finance troubled States.

The Asian Development Bank, the Asian Infrastructure Investment Bank, the Brics Contingent Reserve Arrangement and the New Development Bank (for developing countries, created in 2014), among others, are the large regional banks which aim to promote and finance the infrastructure of their member States without mandatory recourse to the IMF.

The European Investment Back (EIB), created in 1957, borrows on financial markets to finance projects within the European Union and in Mediterranean partner countries.

The European Bank for Reconstruction and Development (EBRD), created in 1990, is an international organization responsible for easing the transition to market economies in central and eastern European countries. It issues bonds on the international market that allow it to co-finance investments. The EBRD's mandate is restricted to countries that "are committed to respecting the principles of multi-party democracy, pluralism and market economics".

The International Bank for Reconstruction and Development (IBRD), founded in 1944 to help rebuild Europe, today focuses its loans and technical assistance on solvent developing countries.

The European Central Bank (ECB), established in 1998, is the monetary institution of the nineteen member States of the European Union who introduced the euro as the single currency. It is responsible for maintaining price stability in the Eurozone and keeping annual inflation "below, but close to, 2%, over the medium term". The ECB plays a central role in supervising the main European banks.

3.1.3. *Technical standardization bodies*

3.1.3.1. *UN/CEFACT*

Topics relating to the Information Society are handled by the UN within CEFACT, the United Nations Centre for Trade Facilitation and Electronic Business. This body ensures that collaboration between governments and companies is maintained in order to guarantee the interoperability of information exchanges between the public and private sectors. It has particularly inspired the EDIFACT standard used in international electronic trade, the architecture for eXtensible Markup Language (XML) electronic commerce and the process used to produce components conforming to the "ebXML" standard, which determines the European framework for IT interoperability relevant to electronic commerce. This "ebXML" standard has also been adopted by China. The other interoperability standard (Dublin Core), which serves the United States as a repository for government metadata records, requires the authorities to implement registration management rules (ISO standard 11179). The European Union does not keep metadata records.

3.1.3.2. *The International Organization for Standardization*

The International Organization for Standardization (ISO) is a non-governmental organization that operates a network of national standardization institutions. Created in 1947, the ISO is an advisory body and its standards are voluntary. The ISO's action areas relate to technical engineering, lasers, the audiovisual sector, acoustics, IT and data processing, cinema and sport, among others. New information and communication technologies are addressed with the International Electrotechnical Commission (IEC), within the framework of the JTC1 committee (ISO/IEC joint technical committee no. 1).

3.1.3.3. *The IEC*

The areas of the IEC, created in 1906, encompass electronics, magnetism and electromagnetism, electroacoustics, multimedia, telecommunications, and energy production and distribution, as well as general related subjects (terminology and symbols), electromagnetic compatibility, metrology, operational security, safety and the environment. Six thousand guidelines have been published.

The IEC cooperates with international and regional standardization organizations, such as the ISO, the ITU, the WHO, the International Labour Office, the United Nations Economic Commission for Europe, the International Council on Large Electric Systems, the International Maritime Organization, the International Organization of Legal Metrology and the Union of the Electricity Industry (EURELECTRIC), among others.

Following a cooperation agreement concluded between the IEC and the ISO, the two organizations created the joint technical committee JTC1/ISO/IEC, responsible for the field of information technology. The leading partner in the IEC is the WTO. This institution has drafted an international agreement on the "technical barriers to trade", which recalls the vital roles played by industry standards and the development of international trade.

The IEC encourages developing countries to take advantage of the benefits that arise from participation in its work. To this end, it maintains close links with the IMF, EBRD, the World Bank and the United Nations Development Programme, among others.

3.1.3.4. *The two third generation partnership project*

The Third Generation Partnership Project (3GPP), created in 1998, seeks to coordinate the standardization bodies in public mobile telecommunications, such as the ITU, European Telecommunications Standards Institute (ETSI) (Europe), ARIB/TTC (Japan), CCSA (China), ATIS (North America) and TTA (South Korea).

It is essential to ensure the highest degree of compatibility between the different radio-based communications systems offered to mobile customers despite the diversity of frequency ranges allocated regionally. The 3GPP ensures that the technical specifications for the mobile standards of public networks are maintained (GSM, GPRS, EDGE, the UMTS and LTE).

A second group, known as 3GPP-2, handles the standards of public, radio-based communications systems in Asian and North American countries, with the assistance of the American standardization group ANSI/TIA/EIA-41.

3.1.3.5. *The International Telecommunication Union (ITU)*

The ITU is the UN body responsible for telecommunications and information and communications technology (ICT). The ITU brings together network owners and the different operators and manufacturers in these fields. It drafts the technical standards and coordinates the regulations relating to the radiocommunications sector (ITU-R) and to networks and fixed equipment (ITU-T). The ITU's third sector (ITU-D) encourages the development of networks and services in less developed countries through training and appropriate financial assistance. The ITU has ties to UN/CEFACT and all technical standardization bodies, as well as the major global financial institutions, such as the World Bank, the IMF and GATT (Appendix 6).

The ITU's Commission on the World Radiocommunications Conference organizes the allocation of the frequency spectrum according to service category every 4 years for all world regions and keeps the relevant radio-based regulations up to date. The ITU allocates radio frequencies and satellite orbits and develops technical standards, which ensure the interconnection of networks and technologies.

3.1.3.6. *ITU-T study group 3*

ITU-T Study Group 3 (economic and policy issues) is responsible for applying the principles of economic policy and the regulations determined by the different bodies under the umbrella of the UN. As such, it carries out the following activities:

– establishing the compatibility and tariffs of international telecommunications services (including methods of calculation) for all fixed-line, satellite and mobile connections;

– establishing accounting regulations on the settlement of accounts;

– establishing tariffs at the lowest rates possible, which are compatible with an efficient service and take into account the need for the independent financial management of telecommunications to have a solid basis;

– organizing global forums to improve understanding of the economic and financial issues related to the growth of ICT, the transition to IP protocols and the development of NGNs;

– coordinating between member States, service providers, member academies and international organizations (including the WTO).

ITU-T SG 3 is made up of several regional groups, which follow the rules formulated by the regional institutions in charge of regulating international services and the major recognized financial organizations [BLA 11].

The specialized working groups are in charge of proposing, after carrying out research, solutions to the various issues presented by the current state of networks, such as:

– international Internet connectivity;

– transition from IPv4 to IPv6;

– international mobile roaming;

– mobile financial services;

– economic impact of over-the-top (OTT);

– governance, etc.

3.1.4. *Globalization and economic growth*

Global economic growth must, in theory, enable production to become specialized and, through the adoption of international standards, trade to increase within a competitive setting. The latter is intended to improve the quality of products and services, reduce costs and encourage technology transfers to the benefit of all.

The "Washington consensus" is a set of financial measures, determined by international financial institutions based in Washington (including the World Bank and the IMF), which must be applied to national economies facing up to their debt. These rules particularly focus on liberalizing foreign trade, eliminating customs barriers, privatizing State companies or monopolies, deregulating markets and protecting intellectual property.

Since 1990, the logic of this economic theory has prevailed and the globalization of the economy has even spread to emerging countries, China and the countries of the former Soviet bloc. However, since 2014 the economic indicators of international trade have stagnated or even experienced a downturn, despite the correct functioning of the abovementioned institutions. Consultation between regional or global financial institutions helps to mitigate crises, with the IMF, for example, providing aid to economies in difficulty, subject to the prior agreement of the ECB or the IMF.

Atmospheres of social or financial crisis are likely to weaken the global balance of peace. The UN's ambitious program of Millennium Development Goals, which should have been achieved by the end of 2015, has not been realized. The disparity between developed countries and developing countries has actually increased in recent years and, in an attempt to rectify that, the organizational arrangements of the structures described here are under review. These international structures are particularly reproached for excessive bureaucracy, the absence of balanced representation and the harmful existence of veto rights.

3.2. The European regulatory framework

3.2.1. *The European Commission*

The European Commission proposes Directives, adopted by the European Parliament, which must then be written into the laws of each member State. The Directorate-General for Communications Networks, Content and Technology (CNECT) and for Competition (COMP) each take action in their respective field.

3.2.2. *BEREC*

Created in 2009, the Body of European Regulators for Electronic Communications (BEREC) replaced the European Regulators Group for electronic communications networks and services. Its work consists of contributing to the harmonious development of the European market for networks and communications services for the benefit of consumers and businesses. Within this framework, BEREC assists the national regulatory authorities of the 28 member States. This institution's lack of adequate power accounts for the gradual weakening of the concept of universal service in Europe.

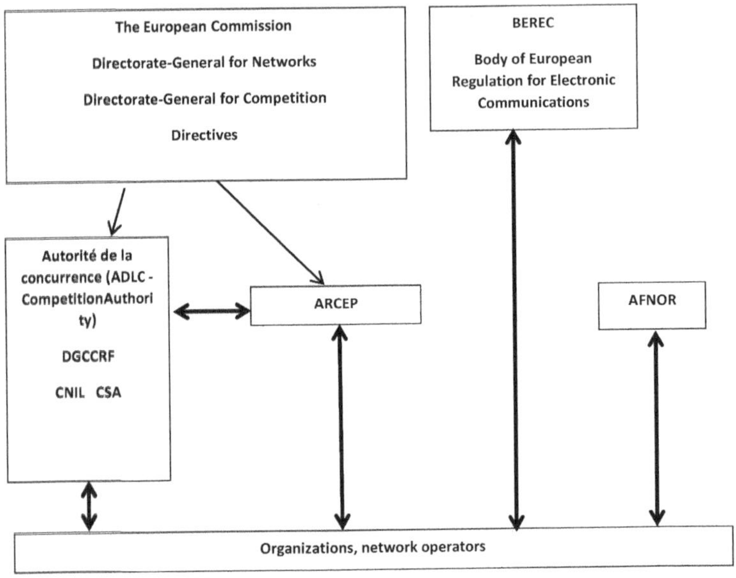

Figure 3.2. *Relationships between European and French bodies*

3.2.3. *Standardization of telecommunications and ICT*

3.2.3.1. *The European standardization framework*

The European establishment endeavors to develop a consistent standardization and regulation system for telecommunications and ICT for the 28 member States of the Union through documents drawn up by the global technical bodies listed in section 3.1.3. The European Standards Organisation system which, in conjunction with the European Commission, manufacturers and SMEs, works to spread standards and European innovation, is made up of CEN, CENELEC and ETSI.

3.2.3.2. *The CEN-CENELEC*

The CEN (from the *French Comité européen pour la normalisation* – European Committee for Standardization), the recognized European standardization authority, deals with domains outside those of CENELEC and the ETSI. It reports to the IEC and is formed of around 30 national standardization bodies from the European Union and associated countries. CEN standards are referred to with the letters ENV.

CENELEC (from the French *Comité européen de normalisation électrotechnique* – European Committee for Electrotechnical Standardization) is a committee established by European Union countries and whose activities correspond to ISO and IEC activities in the field of electrotechnology to guarantee the security and protection of customer environments. The CEN and CENELEC have a common secretariat, CEN-CENELEC.

3.2.3.3. *The ETSI*

The ETSI was created in 1988 to achieve the objectives of the European Union. The ETSI, a non-profit organization, is made up of specialized technical committees and handles partner projects. For some decisions, it operates through voting by national delegations. ETSI only deals with standards relating to systems and equipment and does not handle standards related to connections or networks.

Within ICT, the ETSI contributes to standardizing smart cards for mobile telephony (Appendix 7), intelligent transport, connected homes and the Internet of Things (IoT), among others. The ETSI writes European Technical Standards, Technical Basis for Regulation and Common Technical Regulations standards, the mandatory technical foundations.

3.2.3.4. *The CEPT*

The European Communications Office, the permanent office of CEPT (European Conference of Postal and Telecommunications Administration; acronym from the French *Conférence européenne des Postes et Télécommunications*), supports the Electronic Communications Committee and liaises with the ETSI and the ITU. The functions of the CEPT, which unites 48 European countries, are carried out through the European Public Telecommunications Network Operators for network operation.

3.3. Main French authorities involved

3.3.1. *ADLC*

The French Competition Authority (*Autorité française de la concurrence* – ADLC), created in 1986, handles complaints that are submitted to it and deals with competition law, abuse of dominant positions and agreements, among other things. It has the power to impose sanctions and can take advice from ARCEP (Autorité de régulation des communications électroniques et des postes – Regulatory Authority for Electronic and Postal Communications). Its role is to contribute to improving the competitive functioning of markets.

3.3.2. *ARCEP*

ARCEP implements national regulations on electronic communications and postal matters. It determines sectoral regulations and monitors their enforcement by the Competition Authority. It seeks to resolve disagreements that may arise between companies and also takes action before bids are introduced to regulated markets. The different "national regulatory authorities" are responsible for regulating the telecommunications markets in European Union member states.

3.3.3. *CNIL*

Created in 2004, the National Commission on Informatics and Liberty (*Commission nationale de l'informatique et des libertés* (CNIL)) is the administrative authority responsible for ensuring that public and private IT services do not violate human identity, human rights, privacy or individual or public freedoms. The advice of the CNIL must be sought before any draft bill relating to protecting personal data is referred to Parliament. The CNIL allocates labels to products or procedures on protecting people in relation to the treatment of personal data. The majority of developed countries have each set up a similar institution to the CNIL.

3.3.4. *DGCCRF*

The Directorate-General for Competition, Consumer Affairs and Prevention of Fraud (*Direction générale de la concurrence, de la consommation et de la répression des fraudes* (DGCCRF)) is a service of the French Ministry of the Economy, Industry and the Digital Sector that aims to monitor trading conditions between businesses in order to ensure the transactional loyalty of consumers. Its purpose is to ensure competitive regulation of markets, economic protection and consumer safety. In particular, it takes action on issues relating to online banking, Internet advertising relating to high-return investments, and the ideological aspect of collaborative economics, among other things.

3.3.5. *CSA*

The Superior Council of Audiovisual Content (*Conseil supérieur de l'audiovisuel* (CSA)) seeks to guarantee freedom of audiovisual communication in France. Its responsibilities cover numerous topics, including the protection of minors, respect for the pluralistic expression of opinions, the defense and showcasing of French language and culture, the accessibility of information for persons with an hearing or visual impairment, action to support health protection, the organization of election campaigns on the radio and television, the allocation of frequencies to operators, respect for human dignity and consumer protection, among other things.

3.3.6. *AFNOR*

AFNOR (*Association française pour la normalisation* – French Standardization Association) is the French member of CEN and ISO, as well as of CENELEC and the IEC through the French Electrotechnical Committee, which it hosts. Under the Ministry responsible for industry and in the role of directing and coordinating the French standardization system, which it was assigned in 2009, AFNOR acts as the main organizer of standardization in France, taking note of standardization needs and rallying interested parties. It presents French positions at the European and international levels. It organizes a public consultation on every draft French, European or international standard, approves the final version and incorporates it into the national catalog. It ensures the continuing relevance of the standards published.

The RNF (*Réseau Normalisation et Francophonie* – Network for Standardization and Francophone Countries) association brings together over 75 active members from 27 countries and three continents. Its principal objective is to lead wide-ranging actions to strengthen the capabilities of national standardization organizations and others involved in quality infrastructure, which use broad standards as a development tool. The RNF Association is a partner of AFNOR.

3.4. 1980s tariff principles

3.4.1. *Assessment of implementation costs*

3.4.1.1. *Basic principles*

The basic tariff principles in international communications were established by the ITU-T at the beginning of the 1930s, and then regularly revised (Appendix 8). The year 1988 saw the adoption of the International Telecommunication Regulations, which determine the use of leased lines, essential elements for transporting data communications and establishing future Internet and TCP/IP connections.

For social and political reasons, the tariffs for some services can be established in such a way that they do not cover all of the above-mentioned charges. In addition, the tariffs applied should not trigger harmful competition between different telecommunications services. Taken as a whole, the surplus revenue of telecommunications services should not greatly exceed the amount necessary for the proper functioning of these services (Rec. D.5).

The ITU recommends that international partner operators freely select one of the three methods of invoicing and financial regulation, according to the volume of traffic between the two countries in question and the choices of these countries:

– the accounting rate system, with rates being assessed on the basis of costs by the Regional Tariff Groups of Study Group 3 and taken into account during bilateral negotiations;

– inclusion of all revenue by the initial operator (sender-keep-all) in the case of weak traffic flows;

– flat rate remuneration on the basis of the real usage of the available means, according to the duration and distance in question.

The Regional Tariff Groups of the ITU-T encourage the use of two methods to determine the costs of communications services, with one method based on the cost price, the so-called analytical method, and a synthetic method.

3.4.1.2. *Analytical method*

The analytical method requires precise information on the equipment used for routing traffic and on the actual volume of traffic carried along each direction of the connection. Therefore, for each traffic unit and each connection to another network, every operator should be able to assess through cost accounting:

– the depreciation costs of equipment;

– the financial charges related to the capital invested;

– staff costs and operating expenses;

– consumables required for maintenance;

– the costs of research and trials;

– the taxes and charges related to the work carried out by third parties.

The analytical method allows real cost prices to be fixed at a reasonable rate for each traffic unit on each communications axis. These prices are expressed in a designated international currency, the most common being Gold Francs, SDRs or euros. These reasonable and acceptable values are used for a set period. The ITU provides advice for establishing cost prices, as well as depreciation tables for investments in buildings, cables and equipment. It is accepted that related information should be used and comparisons and parallels should be made in cases where the cost prices are unknown (Rec. D Supplement 1, Cost and Tariff Study Method).

Recommendation D.5 also shows how to evaluate cost prices. On the other hand, Recommendation D.150 underlines the distinction to be made between the accounting rate and the collection charge. International accounting procedures are also explained in detail.

Recommendation D, Supplement 2 provides likely examples of applying the analytical method to assist the Regional Tariff Groups. Recommendation D, Supplement 3 is dedicated to establishing costs and tariffs on a national level. It also explains how to determine the cost and profit centers of network elements, as well as their charging methods (direct costs, indirect costs and acquired costs).

3.4.1.3. *Synthetic method*

The owners and operators concerned focus on the stated average of the accounting rates collected between countries in the region for a relevant service and take on half of this amount, except in cases of transit where the third-party country should receive remuneration for services rendered.

3.4.2. Background to circuit switching

3.4.2.1. *Technical parameters*

From 1930 until the 1980s, a network only consisted of circuits operated manually or automatically. Long-distance analog or digital connections were established with four wires, two for emitting and two for receiving. In each direction, the voice operation of these telephone lines was carried out less than 45% of the time due to alternating conversation. Circuits were rare and demand high, which explained the high tariffs, at times regulated according to day and time depending on the direction required. The structure of networks based on circuit switching helped to establish basic tariffs through the implementation costs of the services offered.

3.4.2.2. *Voice service circuits*

The pricing system of international circuits rests on three factors: the distance between callers, the connection time and the time of day. Rented four-wire lines are suggested to companies wishing to have a secure, permanent connection with a guaranteed level of performance. The basic monthly rental tariff is established on the basis of the price, a switched network, the same connection for 5 h per working day, a figure which, gradually, will be reduced to 3 h, and then finally to 1 h and 30 min.

3.4.2.3. *Data circuits*

Since 1960, price setting has become more complicated due to the availability of modems that, on good quality rented circuits, allowed data transmission at 2.4 then at 9.6 kbps. Taking into account the filtering of analog circuit carriers, data transmission can only be carried out using frequencies of the primary analog group in order to avoid distorting the group's propagation time. This type of limitation required specific management and therefore involved high costs, which disadvantaged ICT development.

Data transmission at higher speeds, for example at 48 kbps, required a complete primary analog group (12 circuits) to be dedicated to it, an expensive process that drew upon resources related to the era's significant voice traffic needs.

Finally, around 1985, the availability of more efficient modems and the gradual spread of digital transmission opened up new paths for data transmission at more acceptable pricing conditions, whether for rented specialized lines or for the networks dedicated to these exchanges. In 1988, private owners and operators were authorized to use rented lines to provide data services, facilitating the expansion of carrier networks of TCP/IP protocols and Internet applications.

3.4.3. *Calculation of accounting rates*

3.4.3.1. *Direct connection*

A network operator does not have the option of establishing an international end-to-end connection. It must negotiate with an authorized partner in the country regarding installation of the means required for connection. The necessary technical coordination is related to sharing the revenue from operating traffic carried on the relevant circuits installed.

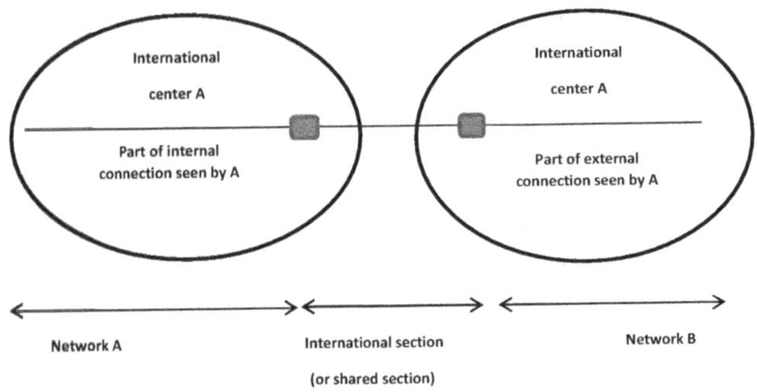

Figure 3.3. *Connections between circuit-switched networks*

Accounting rates concern the taxes collected between operators for the services performed in transit operations or for traffic charges on the final initial or receiving parties. Figure 3.3 shows an AB connection between two subscribers using a direct international connection (without transit country) for 1 min. If "a" and "b" represent the usage cost of the communications resources of countries A and B, respectively, if "c" and "d" represent the transit costs of the international centers in countries A and B, respectively, and if "e" is the costs of the circuit between the two centers, the sum of S is:

$$S = a + b + c + d + e$$

"S" represents the cost of 1 min of connection between two customers. This sum of "S", increased by coefficient "K", determines the rate per minute of a busy hour between the two countries, as settled by the users ($S \times K$ = collection charge).

In connections between international bilateral operators, the proportion of the accounting rate to be repaid to country B for services performed in a call directed from A to B is expressed by the following formula: (0.5 e + d + b) multiplied by the margin of "k". This value, expressed as an international unit of account (for example, in SDR, according to the partners' choice), represents the call termination charge. This is the transmission cost billed by one operator to another operator for carrying out the transit or termination of the call on its network.

In practice, the values of accounting rates are never made public, not due to deliberate impenetrability on the part of the most significant countries in international traffic, but because, devoid of their context, these values have little actual meaning. In fact, they depend on the means used and accounted for, as well as the traffic that has actually been established.

3.4.3.2. *Cases of international transit*

In the event that the necessary connection between countries A and B requires transit through third-party country C, the operator in transit country C should be remunerated and the methods listed in section 3.4.1.1 (detailed breakdown by traffic unit or flat-rate remuneration) should be applied. In circuit switching, no country can set up direct connections with all other countries. Furthermore, each operator has its own list of preferential operators with which it negotiates transit conditions in advance.

In general, operators divide the total of the accounting rates on the basis of the simplest formulas possible, according to the 50/50 return for each of the two terminal parties in the case of direct connections and according to 40/20/40 or 30/40/30 returns depending on the size of the task carried out by the transit country operator.

3.4.4. *Collection charges*

Collection charges are the prices paid by users and are based on a contractual document, which is passed to the customers subscribing to the service. The relationship between accounting rates and collection charges can often seem disproportionate. It can vary, for example, between 2 and 10, or even more. The free discussions undertaken on this topic between experts from different countries over the past 60 years have shown that this relationship depends a large number of factors, such as the youth of the network, the low volume of traffic on some international routes, the difficulty of balancing the network's global budget, the domination of highly influential neighboring countries, the weak purchasing power of a large part of the population, the continuous economic imbalance between different parts of the country and the need to invest in order to improve the capillary network, among other things. The perspective granted by recent history shows that the significant reduction of network costs, which led to the reduction in tariffs and the massive growth in the number of customers, is largely connected to technological progress and investment.

3.4.5. *Network access and use*

Tariffs normally have two elements, one corresponding to network access and the other linked to network use. In the countries of origin, the collection charge used is the, possibly inflated, rate corresponding to the chosen network for the charges collected for use of additional equipment in the country of origin, for network access, or for equipment used in network interoperability.

3.4.6. *Practices in 1985*

Logic would dictate that tariffs were based on the actual use of the equipment set up at the connections described in sections 2.2 and 2.3 depending on the usage time of the different equipment, protocols and software implemented. However, between 1960 and 1990, it was necessary to ensure the consistency of tariffs despite the coexistence of analog and digital technologies.

In the years around 1985, expenditure was organized according to the theoretical distribution shown in Table 3.1 and modified in accordance with local and national contexts.

Expenditure	Percentage
Staff	27
Staggered repayments	32
Loan repayments	11
Consumables	16
Other	11
Profit	3
Total	100

Table 3.1. *Breakdown of a network's management expenditure in 1985*

3.5. 1990s reform

The privatization and liberalization of telecommunications markets was brought about by the agreements on trade in services (GATS) signed in 1994 and 1996. These agreements recommend that tariffs be based on costs and be at a reasonable level associated with transparency in accounting transactions (Table 3.2). WTO member States have adopted a set of additional commitments, acting as a basis for arbitration in case of conflict. The ITU-T, in contrast to WTO and without specialized bodies for dispute resolution, recommends a consensual and flexible approach for payment of services carries out internationally. However, while the ITU is unaffected by the pricing systems of national services, operators are required to have some regulatory and pricing consistency nationally and internationally. On the other hand, international regulatory bodies, including the OECD, advise States to implement control systems to avoid obstacles to the liberalization of services.

Traditional system of levying taxes (accounting rates)	New post-1990 system interconnection rates
Symmetrical in principle (accounting valuation divided by two)	Asymmetrical (costs can vary according to country) If traffic is similar, no payment
Negotiated bilaterally	Determined unilaterally, but subject to trade rules
Discriminatory (rates calculated according to callers)	Non-discriminatory (even the reference interconnection offer available to all network operators)
Half-circuit system	Circuit system (can be unbundled)

Table 3.2. *Development of tariff principles*

3.6. Tariff principles in force in 2016

3.6.1. *NGN technology*

Since the 1990s, international interconnection has taken into account the complexity of NGN, and it follows new rules that have led to significant price reductions, justified by new, less expensive technology and a higher volume of traffic.

As shown in Figures 2.10 and 3.4, NGN includes fixed circuit-switched networks (public-switched telephone networks) inherited from past use and associated with mobile networks, and IP networks, which allow access to the Internet and content delivery networks (CDN) via IMS platforms, with IP transit being carried out "peer-to-peer" (P2P) by IXP/GIX centers answerable to the ISOC.

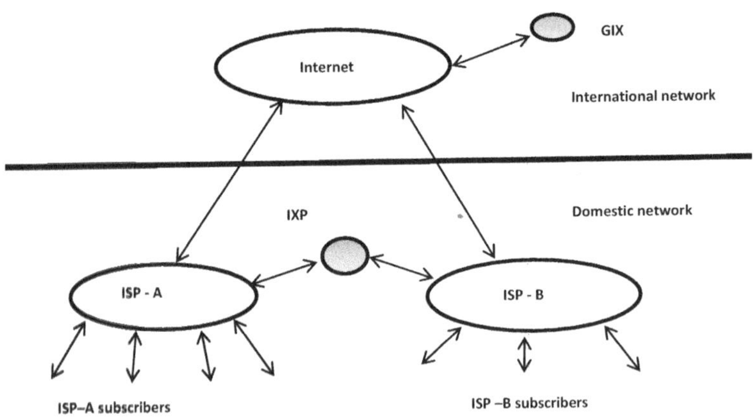

Figure 3.4. *GIX and IXP: Subscribers to Internet service providers A and B have access to the Internet through peer-to-peer GIX and IXP hubs*

The remaining circuit-switched networks are also interconnected to ensure non-IP internetwork connections and, in this instance, the tariff rules in force remain those described in section 3.4. These networks will begin to vanish when the first platforms, the so-called "plaques" or "subregional areas", are shut down between 2016 and 2024. Consequently, operators are offering subscribers solutions with IP connections through ADSL or fiber-optic connections.

The billing and accounting applicable to NGN is determined by Recommendation D.271 which specifically indicates that a session set-up charge can be applied for each session established. This tax can take into account the resources used to set up the session and make it available, such as the transportation and processing (or not) of messages to all relevant hubs on the route, making calculations related to the route, carrying out call access controls for all the hubs involved and the capacity reserved while setting up the session in both directions. Recommendation D.271 also clarifies that, in order to reduce the number of factors used for the accounting procedure among administrations or operators, the parameters of several session can be combined to create a reduced set to which one charge is applied. This merging is carried out over an agreed period – 1 month, for example. This method is often retained for simplicity.

Direct interconnection through IXP/GIX has the advantages of reducing costs and latency time, and being able to guarantee availability in the data rate. Traffic exchanges made within GIX are carried out without additional cost. The traffic passing through an IXP is not charged, although the flows toward the traffic provider from the ISP are charged. IXP/GIX also route the traffic from CDN.

3.6.2. *Internet traffic*

3.6.2.1. *Bilateral agreements*

In December 1998, the revision of Recommendation D.150, regarding remuneration procedures for international traffic routed to end parties, was approved. The accepted principle is of one rate reflecting the costs in cases of asymmetric traffic for the signatory countries of international agreements on trade liberalization.

3.6.2.2. *Internet interconnection*

ITU-T Recommendation D.50 recognizes the sovereign right of every state to regulate its telecommunications and to negotiate and conclude bilateral trade agreements. The objective is to establish direct international Internet connections, which take into account the potential need for compensation among organizations, managers or operators, regarding the value of elements such as traffic flow, the number of routing channels, geographic coverage and the cost of international transmission.

Interconnection modalities, and consequently taxation methods, may include the following: P2P links, transit links, hybrid forms (P2P or transit links) and any provisions agreed through common agreements between organizations (or managers or operators), including indirect interconnection, as envisaged within the Global Information Infrastructure (GII – Rec. Y.140).

Respecting the rules of neutrality, traffic exceeding network capacities can cause indiscriminate slowdowns in data transmission, or even a complete freeze. This sometimes occurs when problems with equipment occur on the key transmission routes, for example. Routing regulations then serve to split the traffic load across the nearby networks according to their capacity. At the national and regional levels, the implementation of Internet exchange points (IXP/GIX) and content replication systems (caching) for the most frequently requested pages should be organized.

3.6.3. *Mobile traffic*

3.6.3.1. *Principles (Rec. D.98 and D.99)*

The wholesale and retail roaming tariffs (see section 2.2.1) correspond to the prices charged by the international mobile roaming (IMR) service, for instance:

– IMR's wholesale tariffs match the prices that the operator of the network visited has charged the connecting network, so that the customer could access the operator's network from the network visited;

– IMR's retail tariffs correspond to those applied by the operator of the connecting network for subscribers to IMR services. Regulatory bodies can take action on the issue of international mobile roaming tariffs for the benefit of users by encouraging competition;

3.6.3.2. *Toward "reasonable roaming"*

Due to the high cost of the initial investments required for parts of fixed-line and wireless networks, tariffs have always been high. In the absence of regulations, when operation of voice GSM began, it was decided that callers would be made to pay, with the party being called paying nothing, even though in the previous system (RadioCom 2000) the mobile subscriber being called was charged. On this issue, Europe could imitate North American practices where the mobile device called was charged, as well as the radio party (FCC decision). High tariffs have enabled the rapid development of mobile networks.

Since 2012, the strong growth of mobile networks in numerous countries has put mobile telephone networks in a position that could be described as "dominant" in some markets. With monopolistic attitudes undesirable, the guideline tax for international termination on mobile networks should become closer to that of fixed-line networks. For this reason, the European Parliament and the European Commission have proposed that the roaming charges paid by mobile network operators in their 28 member States be abolished. Mobile users would therefore pay the same price for voice calls, SMS or data transfers, wherever they were in Europe. All additional roaming charges will be banned from June 15 2017.

The abolition of roaming charges has become a timely and justifiable action. On May 1 2016, the tariffs for calling a mobile device from abroad with a national subscription will decrease (€0.05/min, €0.02 for SMS and€0.05/Mbps for data). On June 15, 2017, these charges will be abolished within the limits of a "reasonable" usage yet to be defined. The purpose of the IoT does not yet seem to have been resolved as network managers and operators are invited to share their resources. In this case, with regard to roaming, the IoT has not brought about repayments between operators as the calls in question are indistinguishable from ordinary calls. Official implementation documents should soon be finalized.

The high cost of mobile roaming is a topic that also concerns developing countries. Eastern and African countries have created the

"One Network Area", with its aim being to markedly reduce the costs of roaming services in mobile telephony.

3.6.4. *Current practices in 2015*

3.6.4.1. *Overview*

Over half of the traffic routed though Internet access networks in France is delivered by just five independent systems (across the 70,000 networks that make up the Internet), corresponding to service providers, hosts, or technical intermediaries.

As of 2016, modifications to systems or standards require equipment to be renewed every 5 years. Next generation networks are relatively complicated and it would be difficult to accurately add up the different elements used. Furthermore, network operators regularly carry out cost assessments of the global holdings they require for the life of the network, adjusted by fiscal bodies and the pressures exerted by competition.

3.6.4.2. *Mobile data*

In the case of mobile transmission of data, the caller and recipient each pay a share depending on their tariff plan, but aspects of roaming produce the same results as with voice data (low rates within Europe, high rates outside Europe).

In the IoT, it is even more beneficial to be roaming as the regulations oblige mobile operators to share their networks, allowing complete and continuous coverage at the lowest possible price.

3.6.4.3. *On-net and off-net traffic*

An "on-net" call is one that terminates on the network of the initial operator. An "off-net" call terminates on the network of an operator other than the initial operator. The tariff differentiation between "on-net" (intranetwork) and "off-net" (internetwork) can be a source of contention between operators. Exchanges of equally important traffic between two partners do not result in payment. If the data feeds exchanged are of different sizes, the compensation required is assessed on the basis of wholesale prices, with the gain (or loss) for each operator arising from the price gap.

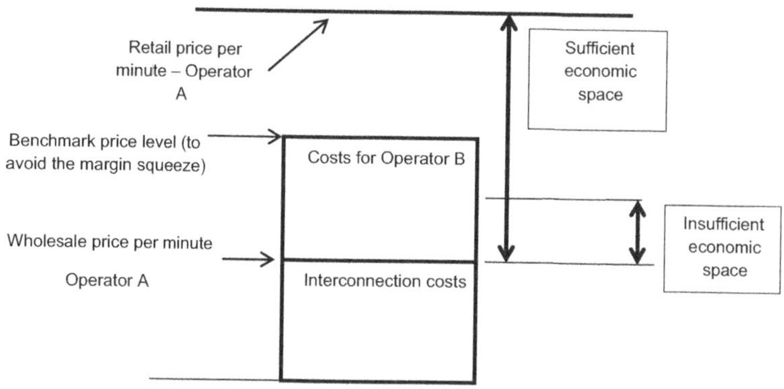

Figure 3.5. *Pricing stances of operator offers (according to TERAConsultants)*

A "margin squeeze" occurs when the gap ("economic space") between the retail price of operator A and its wholesale price is too small for operator B to be able to compete with A in retail. The competitor's bid is therefore not replicable.

The gap between "on-net" and "off-net" tariffs should not exceed the gap between the associated costs in order to avoid the differentiated offers being seen as an anticompetition tactic. The tariff differentiation can be abused by dominant operators to restrict the volume of off-net calls and make small operators less attractive, leading to a reduction in the level of customer satisfaction. Forbidding tariff discrimination beyond the level justified by cost difference puts an end to this source of market power (Figure 3.5).

Through "asymmetrical call termination" in IP, the operator of the user transmitting the call has the opportunity to increase its profit (sometimes to the detriment of the data rate and the quality of service). It is often possible to terminate a call in a neighboring country with cross-border traffic on the network where the subscriber is located, with the assistance of equipment sold freely on the Internet. While regulators can engage with network operators in the wholesale market, it is difficult for them to do so in the retail market as they have neither all the necessary information nor the means to act in a timely manner.

An inadequate pricing system for call terminations is likely to artificially restrict the capabilities of the operators with the weakest parts of the market to participate competitively. Indeed, operators with the strongest parts of the market can "free themselves" from the cost of terminating a mobile call by creating generous offers limited to single calls to their own customers (generous offers benefiting the network). The operators in the weak part of the market are punished when the mobile termination rate is a fixed at too low a level in relation to the real costs of mobile networks. The regulations regarding the retail price are difficult to apply, as an operator may meet the market's lowest pricing conditions for its customers while gaining 75% of the market and advancing its EBITDA. The "dominant" position here is simply a reflection of typical business acumen and it does not lead to a "monopolistic" action. The contradiction is a matter of how the regulations are interpreted.

3.6.4.4. *P2P*

The volume of exchanges between "peer to peer" users is impossible to measure and is therefore not chargeable (see section 2.7.2 and 4.4).

3.6.4.5. *OTT*

OTT actors (including Facebook and YouTube) enjoy very significant advantages as their traffic is asymmetrical and they do not pay for the large amounts of data sent or received beyond the connecting fiber-optics. It is possible for the user to circumvent high costs by means of OTT applications (with the use of voice over IP (VoIP) on WhatsApp, Viber, Libon and Skype, among others) using Wi-Fi when appropriate.

3.6.5. *Conclusions*

It is possible for a user to circumvent high costs by means of applications using VoIP or P2P technology, or through applications offered by OTT (WhatsApp, Viber, Libon and Skype) and using Wi-Fi when necessary.

The solution is that international traffic passes through all IP, which is much less expensive than circuit switching in terms of equipment, provided that there is a data rate of a suitable quality, appropriately sized international connection points and properly negotiated tariffs.

3.7. Pricing practices

3.7.1. *Tariff levels*

Different approaches depend on the situation of the market. Tariff levels should be maintained at an acceptable value by the relevant part of the market so as to create customer loyalty, and to avoid provoking unhealthy competition with competitors or related services.

Incidentally, if tariffs are too high, customers are driven away from the services and limit their communications budgets, which can justify, with the involvement of the regulatory authority, the introduction of a new operator. Conversely, too much competition destroys the market through an inadequate quality of service. In the liberal economy of 2016, this theory is not restricted to the law of supply and demand, but it suggests that innovation goes hand in hand with competition, without specifying the desired outcome and with the door thereby open to other abuses.

The topic of pricing systems resembles a never-ending game in which the seller tries to ensnare the user and where the user simultaneously seeks to avoid the trap of high prices through offers from competitors or new devices. When a suitable agreement cannot be found, the game has two losers. However, it turns out that the network operator is acting in tandem in two complicated games – one with its customers and the other with international partners and industrial suppliers. The outcome is never clear and the current situation (an international economy and constant technological renewal) makes the exercise ever more complicated.

In Hungary in 2014, a draft tax on Internet usage was proposed in order to help finance national industry. The proposed tariff, roughly 0.47 euros per Gb, would in principle allow 90 billion euros to be collected, but the public immediately protested to show its disapproval. Indeed, this tax would have reduced traffic enormously and, as a result, would have been profoundly negative for the country's economic development. Furthermore, reducing Internet exchanges would also lead to restricted access to information and freedom of expression being obstructed.

In a nationally sized network, the relationship between the number of professional and residential customers depends on the quality of overall subscriber service. An underdeveloped network serves the majority of public and business services, while a very dense network only holds 15–30% of professional customers. This point is very important as the customer categories mentioned here have very different network practices in relation to the hours of the day and the days of the week. Yet, a network must be developed to handle these two types of traffic without having to allocate routing priorities. There is no model worth keeping on dividing the number of subscribers by category, any more than there was originally regarding the revenue from these categories. Each national network is in some way a reflection of its economic activity and the sociocultural habits of its citizens. Several research units and international organizations have sought unsuccessfully to make connections between the various statistics provided by states, and this research, which was hoped to be highly useful in drawing up development plans, has so far been in vain.

Over the last 30 years, the burden on staff has been considerably reduced. On the one hand, there has been a reduction in the useful financial life of network equipment, which has greatly relieved the financial burden of "depreciation" positions to be taken into account in an operator's financial performance. Indeed, a shorter operating life for equipment requires more frequent renewals and leads to higher repayments. Reducing customer prices is only possible if the traffic increases and the number of customers grows. Switching equipment had an operating life of 25–30 years in the 1950s, which had become 20 years in 1975 and then 12 years in 1990.

Tariffs are the reflection, through products or services, of the needs required by the budgetary statement of revenue. The ITU supports revenue being used to cover the three categories of expenditure:

– network and service usage that the operator makes available to the user. The amount necessary for covering this expenditure reflects the cost and quality of company management, as well as the degree of organization and productivity. This amount can be obtained as the primary outcome of technical and economic research;

– development activities and research that the operator undertakes to ensure its future;

– positions on taxation and other charges, according to the existing system. Communications services can be used to finance ICT in schools, or for space exploration, distance learning or financing parts of a national sector, among other things;

In a liberal economy, commercial action is based on the related tariff levels and sales prices. Tariff modulation according to hours and days of the week allows a better use of network resources by dividing them according to time, for example for business traffic and residential traffic (See Appendices 2 and 3 on pricing experiments carried out in Europe in the 1980s).

3.7.2. *International balance sheet*

3.7.2.1. *Principles and their application*

The practice of exchanging accounts between organizations or operators for international traffic is as old as the networks themselves. Payments are made through a settlement agent in order to take into account the monthly fluctuation in exchange rates and a regional clearing house is responsible for making the annual balance sheets for each operator involved.

Opacity has always prevailed in the relationships between accounting rates and collection charges, mainly due to countries with high traffic and dominant economies. No key stakeholder wants to divulge what it considers to be strategic accounting secrets. This attitude is understandable, although there is evidence of glaring mistakes among the minor open secrets.

In principle, the "account balance" between two operators is calculated according to the results registered by both partners on a monthly or quarterly basis.

The price of call termination plays a key role in the payment systems of mobile operators. It requires thorough accounts management. The accounting revenue division procedure is based on the equal division of accounting rates, a principle defined in the ITU-T Recommendations but one which is not mandatory. The general rule is that, if the international traffic between two operators is balanced, there is no reason to proceed to a financial exchange. On the other hand, a more significant rate of outgoing traffic gives rise to a differential payment.

Offsetting the balances of international telecommunications accounts in a deficit relationship (Rec. D. 196) is carried out with the agreement of the relevant organizations or operators, whether with recourse to a third party country or the balances of international accounts (clearing) in an accounting policy sometimes referred to as relating to "invisible" parties, as its compatibility is not supported by any tangible material (http://www.itu.int/ITU-D/finance/work-cost tariffs/publications/economic_study_for_financing_telcom_dev-don_ monk.pdf).

3.7.2.2. *International switched networks*

Generally speaking, the volume of national switched French-language voice traffic in France is close to 95%. This proportion, which was formerly around 3% in duration, is now currently in the region of 5% of traditional telephone networks. Unlike English, the French language does not have the same sized audience internationally. On the one hand, France is a large country and national uses have naturally come to dominate those uses established in international connections. In comparison, small countries (Brunei, Jordan and Luxembourg) have very large volumes of international traffic. Furthermore, for France, the financial impact of international communications tariffs on the entirety of revenue has always been insignificant, regardless of what some commentators, who do not always have all of the accounting information, may have said on the subject.

In reality, considered together with their connections, French international switching centers are falling ever more into deficit as their traffic is split into a large number of connections. Only around 10% of international connections are economically profitable and the sum of these profits does not manage to compensate for the shortfall from other connections, which are most often only maintained for reasons of prestige. Packet technologies and the use of IP routers have completely changed these facts.

3.7.2.3. *International switching in packet mode*

Needless to say, as far as packet mode on the Internet is concerned, the situation has changed completely. Indeed, the technology puts messages on P2P connections geared toward specialized IXP and the real costs of electronic communication have become very low, so that the settlement of accounts operates on the basis of traffic volume predictions, which are often the same or similar in both directions of transmission. If this is not the case, redirecting traffic to other IXP must be considered to move closer to a balanced situation.

In 2010 and 2012, the European Union stated that the international communications made with mobile devices represented a marginal share of the total (between 1 and 2% of mobile telephony revenue globally and 4% of community traffic).

For its part, China recognizes that if the development of OTT services on NGN has a significant and lasting impact on ICT development, these OTT services will have a negative effect on the revenue of traditional operators of voice and data services. Several countries wish to resolve issues linked to competition from OTT services, in particular those connected to VoIP services by implementing international regulations, which are more favorable to national operators. The mandatory participation of OTT services in national network investments is often mentioned.

It should be noted that there is no universal rule that can be applied to the pricing systems of telecommunications services. There is only a multitude of specific cases that it is impossible to analyze in terms of the network's years of service, the digital significance of the number of subscribers, their profiles, the economic possibilities and the predominant sociocultural factors.

3.7.3. *Sales revenue and GDP*

In general, the sales revenue in a country's telecommunications services is expressed by a value close to 2% in relation to the GDP of the years 1989–1991 (between 1.70 and 2.20), with some extreme values of 2.55 and 2.42 for Australia and the United Kingdom. According to ARCEP, in 2011 all telecommunications and ICT activities taken together reached a total of 100 billion euros (5% of GDP), with the entire electronic communications services market increasing to 41 billion euros (or 2.05% of GDP). The OECD estimates that the telecommunications revenue of countries has varied between 2 and 3% of the value of their GDP between 1985 and 2007 [OEC 09].

Between 1998 and 2001, the French telecommunications market increased in value by 85% and in volume by 110%, prompting a 15% reduction in prices in 12 years [ARC 12a].

Significant investments (tunnel construction, canal digging, building and maintaining embankments, substantial civil engineering works) are never reimbursed on a par with the efforts made. As happens quite often, they are the product of multiservice infrastructure works carried out for the profit of social utilities and whose use will benefit a multitude of users over several decades, with no organization able to position itself to take over or to act as the manager responsible for these holdings. Concerning the "depreciation" of a network, its "estimated operating life" (that is taken into consideration in accounting) plays a key role. The cost price of a communications satellite or submarine link, or necessary civil engineering works in the mountains to establish international connections, weighs very heavily in the budget of a small network and the trend should lead to these charges being shared over the largest possible number of years to avoid raising tariffs too much (see Appendix 9).

The quandary, for the operator, lies in providing communications services at the lowest prices, while also being able to repay existing loans and maintain sufficient financial reserves to be able to guarantee its future.

4

Supply and Demand in Communications

Network operators provide architecture and access to next-generation networks (NGN). Different types of communication are available on the Internet and financial services consider this an opportunity to make their applications more varied. The range of applications offered by multiple players is supplemented by numerous offers.

4.1. Providers and customers

4.1.1. *Service provisions*

4.1.1.1. *Overview*

The communications services market cannot be considered a market that is freely open to the choices of users. The availability of all technologies everywhere at the lowest price is a natural obstacle.

Since the creation of national telecommunications networks in around 1890, technical adjustments to the required equipment have been gradually superseded by a pattern in which new generations of equipment are introduced every decade, defined and approved by international standardization. While there have been clear improvements in the provision and quality of service, the rationing imposed by the high cost prices of equipment and network management has fortunately been reduced (Appendix 10).

Faced with innovations, user demand is never expressed clearly and completely. On the other hand, development centers work on the varied concerns of users with relatively long-term goals. They often prioritize technical performance requirements to the detriment of ergonomics adapted to the overly high number of profiles presented by customers. The overview of the main communications services given in the following shows that, in the current state of technology, the confluence of supply and demand can create disagreements of varying levels of significance.

4.1.1.2. *Voice services*

For over a century, technological developments have allowed the quality of voice signal transmissions to improve. It seems that the technology has now reached an optimum technique, which enjoys a global consensus. International agreements have stabilized the definition of the band of voice frequencies transmitted around two definitions; the 300–3,400 Hz range (instead of 200–2,100 Hz for the American speech of the 1930s) and the 150 Hz to 7 kHz range for enhanced bandwidth (known as HD, high definition), which should become essential in the future.

In digital technology, several coding standards are in use in wired and radio-based networks (VoIP). The different versions of coding standards and their shared conversions seem to generally satisfy customers overall, and anything that remains hidden to users belongs to the operators' own area. On this topic, the choice of conversion systems for standards in voice systems is a problem for operators, given the real costs of installing equipment in all network centers. For example, in recent years, many operators have hesitated to set up the VoIP on LTE option (VoLTE) because it has been relatively expensive in comparison to other simpler solutions, and demand has remained low (section 5.4.5.1) [BAT 14]. Indeed, LTE began implementing VoIP in packet mode on radio networks at a time when all operators were still using circuit-switching systems for voice services.

For economic reasons, the logical development would be to only retain the wideband HD voice standard in networks and to reduce the number of options used in radio systems. In general, users are currently satisfied by the sound quality they receive and their requests largely focus on the prices to be paid (fees relating to the terminal, to subscriptions and to the rate of usage).

4.1.1.3. *Written services*

For many years, the transmission of written messages was subject to a particular treatment in networks, as demand was at once very low and completely essential to some clients, that is for the State, the original owner of all aspects of the network, and for those in charge of industry and trade.

From the outset, in 1880, a specific network was created for telegraphy, which is divided into several stages within the Telex automatic network, reserved for State services and businesses (1946), and within a network open to the public with the telegraph service delivering sealed messages (blues) to homes. Despite the existence of different national standards, various efforts were made by administrations between 1970 and 1980 to increase the speed of transmission (telegraphy service at 200 symbols/sec; Teletex service at 2.4 kbps). Due to the high prices of terminals and subscriptions, these commercial practices were never particularly successful. The demand did not respond to the supply offered.

Nevertheless, in 1977, a study on the standardization of Telex in university libraries was carried out. It underlined the need to plan for "the extension of computer networks". Telex's half-duplex mode of transmission between the caller and the party called likely inspired the development of the Minitel terminal (1982), which was very warmly received by the public. Indeed, 9 million Minitel devices were in use in 2000, in comparison with the peak figure of 134,000 Telex terminals recorded in France in 1986.

The major technological innovation, which weakened the Telex network, was certainly fax. Created in 1848 as "automatic telegraphy" and successively modernized in 1920 and 1970 through the availability of new components, fax was a major development for the unilateral transfer and reproduction of hand-written or typed pages in black and white with a medium-quality definition. This service was offered to business users in various forms (Publifax, Viafax and HomeFax). The million and a half fax machines in use in France in 1993 have been gradually replaced by file transfer applications and computer messaging services on the Internet.

Some ill-fated offers have been made to business users, such as Group 3 fax machines, which were operationally incompatible with PBX branch exchanges on ISDN. Group 4 fax machines were largely unsuccessful, as was teletext, designed to become a "super telex". However, fax machines that provide verification of each of the pages transmitted are still in service for some dedicated applications (transmission of files between the judiciary and attorneys, for example).

4.1.1.4. *Data services*

Data services have long been guaranteed through leased lines directly connecting the two distant services involved with the aid of modems adapted to the features of these connections. Businesses, particularly those in the written press, were affected by the high rental costs for these lines (around 100,000 lines with data rates between 9.6 and 48 kbps were rented in France in 1980). Lacking adequate network knowledge, many businesses were surprised by the digital connection offers made by operators and "expert" consultants have not always known how to reconcile the features of leased lines or the available network protocols (frame relays, ISDN, X.25, ATM) with the technical and economic priorities of businesses.

With technology assisting the development of offers, European organizations have suggested constructing national data networks on which businesses could connect. Circuit-switching (Datex-L in Germany) and packet-switching (Datex-P in Germany and Transpac X.25 in France) data networks were then conceived and constructed, with the latter model proving to be the least expensive and most profitable as it is adapted to the data rates of all terminals.

The Internet, the network of networks, has found itself in the historical path of the development of data networks in packet mode. It has currently taken on the transportation of all data services, along with the advantages and disadvantages of its technology and its own features. In return, although Transpac still served 90,000 businesses for their banking transactions in 2011, it closed in 2012 in the face of competition from the Internet.

4.1.1.5. *Image services*

Image transmission began in 1914 with the Belinograph, which, through the possibility of reproducing images on paper, served the press around the world very well until the development of thermal paper in 1970. The broadcasting of televisual images, which began in 1931, only took off commercially in 1950 and the technical developments in this field have only continued to increase. The most recent developments principally concern the variety of display technologies and image formats.

Industry competition has led to the diversification of solutions related to the quality of the image displayed (4 K screen to 4,000 pixels on the screen's horizontal line), without a significant increase in cost price and with an improvement in terminal ergonomics. Several standards have been drafted on the topic of subtitling television images, but few broadcasters are able to offer this facility due to the too great diversity of options in the equipment used in the broadcasting chain.

Network operators are considering the need to broaden their service provisions in broadcasting television programs and retransmitting them to other networks, including the Internet, for example. The image quality depends on the availability of high bitrates at the time of the broadcast. It appears that the only successful examples of media/telecom integration are cases where the operator has been able to retain exclusivity of premium content for its subscribers, with the agreement of the regulatory authority. In France, ambitions to acquire television sporting rights have been completely restricted.

4.1.1.6. *Services for transferring funds*

Having been in charge of transferring funds since the Middle Ages, postal services were all designed to cooperate with telecommunications services for this purpose. Telegraphic mandates, established in 1868 and boosted in 1879 by the merger of these two administrations, used the telegraph and the telephone, followed by the telex and the fax machine. With regard to the post office, a promissory note is a “mandate” issued so that a transfer of funds may be carried out, with the service being paid for through commission paid by the sender, including the transaction insurance costs and potentially the financial exchange rates (currency conversion).

The electronic transfer of funds with a memory card is the result of a long succession of patents filed between 1968 and 1978, which led to the distribution of the Télécarte in France (1983), intended to pay for telephone calls from a public phone booth. Banking cards appeared in 1998 through Gemplus and Gemalto. In 2016, remote payments are possible through cards with EMV (Europay Mastercard Visa) microprocessors used for banking applications with the transmission of a personal code word in client/server communications systems. In this case, it is again the bank that issues the card or begins the service that handles transaction security through the network, which deducts fees and possible exchange rates (see section 4.3.2.2).

Since 2013, with the aim of reducing network transit fees, various offices have been offering their cooperation in order to carry out financial transfers using virtual currencies (Bitcoin, for example) created through P2P (peer-to-peer) software downloaded onto the computers of both relevant parties (see section 4.4).

4.1.1.7. *The balance sheet of supply and demand*

Several common points can be seen in this brief overview of the development of the primary provisions of communications services:

– in a little under a century, the successive stages of technological development have allowed there to be a response to the requirements brought about by a demand that was undoubtedly greater, and therefore more profitable, than could have been foreseen;

– the availability of good quality communications services throughout the national territory facilitates economic development and the spread of culture, without any specific metrics being able to offer support in assessing the efficiency of the factors under consideration;

– the business market is only a small part of national traffic (between 5 and 30% depending on the country) and yet it is this market that accounts for the recently agreed network developments on an enlargement for all users;

– the availability of new components has led to the development of new technologies;

– the need for operators to turn toward new network protocols remains completely impenetrable to users. This attitude is explained by the desire not to confuse the public with a complicated operating mode;

– the corresponding protocols and equipment set-up on a network are not always used to their full potential by customers;

– network costs remain even higher in investment and operation/maintenance than the networks' density;

– the quality of service and the security of transactions require very high cost prices, the reasons for which are not understood by the public;

– customers always experience great difficulty in expressing precisely which service features they would like to have, even in the short term;

– the economic decisions of international institutions have led to the adoption of a digital agenda for Europe, which aims to set up access at 100 for 50% of the European population and very high speeds for all (higher than 30 Mbps) in 2020.

Between 1920 and 1980, businesses that wanted to have their own connections could lease fixed P2P lines, which complied with international standards. These provisions, charged per day or per month, were quite expensive and could represent the price of several hours of continuous connection per working day. The key interest in this offer of leased lines could be considered very differently by user companies: permanent availability of connection, capability to transmit music in HD or data at 9.6 kbps, guaranteed assistance in the event of line breakage and so on. The creation of a data network allowed this system of several tens of thousands of connections, each difficult to manage and highly expensive, to be greatly reduced. Today, technological advances have allowed costs to be reduced while proposing new solutions, which are likely better suited to demand.

In total, in the past as in the present, customers use the service provisions that are available and standardized, with the differences between operators only affecting the pricing level in accordance with frequency of use. Service provisions can sometimes surpass the level of demand, which gives the customer the opportunity to "explore" the range of offers. On a practical level, the communications services market currently finds itself quite far from the theoretical framework of "supply and demand", with the supply being overabundant, in Europe at least, and the demand not always managing to express its own needs.

4.1.2. *Satisfaction of needs in communications services*

The concept of supply, economically speaking, indicates the available quantity of products and services that are ready to be sold. The concept cannot be separated from demand, which is the quantity of products or services that consumers are willing to purchase. The confrontation between these two concepts shows the state of the market for a given product or service, and leads to prices being set.

According to the content of economic and political international agreements, which have been concluded, national regulatory bodies are tasked with watching for any abuse of dominant positions between market players and controlling the "fair and reasonable pricing level of communications services". In these conditions, network operators are obliged to adjust their offers according to the most obvious needs of their customers, which is not always consistent with high profitability of the available resources. Competition between operators leads them to a certain standardization of their offers, sometimes to the detriment of research and innovation.

As an example, one French scientific research body constructed its own computer services, adapted to its own activities and security constraints. This service provision is available as follows:

– *secure Cloud*: virtual servers on demand and according to requirements, secure hosting of websites, bulk data storage and automatic saving of workstations;

– *integrated messaging*: modern features, shared, secured and integrated into other daily tools for greater ease and performance;

– *portal on a secure collaborative platform*: dedicated to researchers, managers and their external partners, as well as administrative and operational bodies;

– *sharing and nomadism*: synchronized multistation access to personalized files and secure "Dropbox" sharing;

– *individual videoconferencing*: from workstations or mobile devices, with interactive document sharing and display management.

In this way, the computer service of this center of activity has been "custom" made, and for a specific purpose, with the assistance of software adapted to its own needs and in conjunction with the connections available on the Internet. Subject to proper coordination with the company's Director of Computer Services, the task of the network operator is reduced to a minimum. The cost of accomplishing this is likely to be quite high and the experience gained remains within the company. The relative levels of security obtained in these two particular cases cannot be compared.

This example is intended to illustrate that balancing needs in communications services can only be custom made through the Director of Computer Services of the company itself. The optimum communications system in a hospital, a technical research center or a large retail space can only be best planned and achieved by the services of a company's Director of Computer Services. The link between a company's IT and its communications requires close collaboration between the company and the network access provider.

4.1.3. *The long tail*

The concept of the long tail was introduced in 2004 by Chris Anderson, an American economic journalist [AND 06]. It concerns the statistical distribution of sales of a product or service. In a way, this expression shows that sales can be considered in two categories; one focusing on a highly responsive but small number of customers (the left part of the diagram), and the other, greater in number, but moving more slowly to a total amount of purchases, which is higher than those of the first group (the right part). Manufacturers must therefore decide, from the conception of their product or service, which part of the market they wish to interest in principle. Customers faithful to a brand and technological devotees are found in the first part of the curve, while the "followers" watch the market and make their purchases through different means over a longer period of time. These two types of sales require different approaches and distribution methods.

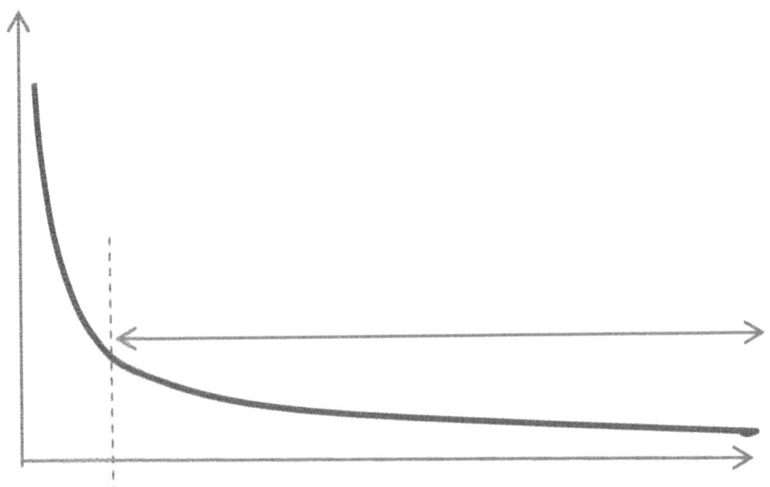

Figure 4.1. *Probability of sales distribution*

Chris Anderson believes that some low demand products can collectively form a larger part of the market than best sellers, if the suitable distribution channels are able to make them known. When the cost of storage and distribution is low, marketing expenditure is also reduced. If, on the other hand, logistical costs are high, marketing cannot last very long. This concept, which supports sales systems on the Internet where the connection, referencing and dynamism of search engines are key, has been taken over by "barbarians" and will be mentioned later (sections 4.5 and 5.3.3.2).

4.1.4. *Monopoly, duopoly or competition*

According to the current international rules, competition is essential to market dynamics and a vibrant commercial atmosphere. As of 2016, the European Union still has 155 network operators and reconciliation operations are underway in Germany, Ireland and even in the United Kingdom, in the hope of reducing this number to 140, which is still much too high in comparison with the United States or China. It is easier to invest when there are fewer competitors. Yet, it is impossible to reconcile investment or innovation with pricing levels that are too low. The European communications services market is too fragmented, being led by precarious players who tomorrow could find themselves dependent on non-European players. At the beginning of the 1980s, the European Commission wished to set up, at the European level, one or two large network operators with a European (or global) dimension. It is hoped that new positive developments could soon occur to allow the European Commission to achieve this objective.

4.1.5. *Billing the data rate*

Technically, it is possible to increase the flow of a network's incoming and outgoing connections, as well as the traffic's main features. The billing and regulation of these parameters establish the basic rules. Each of the applications currently available on the Internet responds to the typical parameters, and billing the data rate would signify the penalization of some applications for the benefit of others and the end of the concept of "Net neutrality" (see section 4.2.5).

> "Over half of the traffic routed through Internet access networks in France is issued by just five independent systems, corresponding to service providers, hosts, or technical intermediaries, across the 70,000 which make up the Internet. The partner at the direct traffic's origin can be identified, unless it acts under the name of a third party. The level of P2P traffic cannot be assessed" (Arcep).

Within the current framework, the measurement of real traffic cannot be arranged and a pricing system according to volume would risk penalizing the small players in the digital economy.

4.2. Obligations of network operators

4.2.1. *Responsibilities of a country's main operator*

In every country, the State must rely on an operator for the necessary technical resources to ensure proper usage of its communications service and, furthermore, if there is no other option, to support its economy. A country's main operator is subject, in principle, to clear obligations, including:

– tariff obligations related to geographic averaging;

– the obligation to offer reduced tariffs to the disabled or people in need;

– the obligation to have payphones for public use;

– the obligation to provide an information service and a directory in both printed and electronic forms;

– the obligation to ensure universal public service;

– monitoring of international standardization;

– organizing vocational training in communications systems.

All of these obligations have a cost, even when they sometimes offer a slight competitive advantage. This feature must be taken into consideration in financial balance sheets.

4.2.2. *Public service tasks and universal service*

Communications regulations cover a set of measures whose aim lies in the desire to protect the consumer. These measures have evolved over time. In the 1950s, the public service tasks entrusted to the French telecommunications service included universal service (complete service for the territory), continuous service (continuous availability of staff in the event of a serious occurrence) and access to emergency services. In 1996, these tasks were redefined and the operator Orange was asked to adopt a limited universal service for fixed-line telephony only (mobile telephony was not included), which focused on the following points:

– connection and fixed telephone service at an "affordable" rate for voice, fax and reduced rate data usages;

– management of an information service (118 612 in France);

– providing a departmental directory;

– access to emergency telephone services (firefighters, police, etc.);

– providing dedicated services and facilities for disabled people;

– installation and maintenance of payphones for public use (presently limited in number due to the development of mobile networks).

As of 2016, in most countries all of these considerations have altered further due to technological development and uses. Payphones should have disappeared by the end of 2017. Two arguments have been put forward to justify this decision in France: maintenance costs, which are close to 8 million euros per year, and an average usage, which in 2014 had gone down to < 1 min/day and per booth.

In France, public service activities in telecommunications, intended to serve the public interest, fall into the following two categories:

– state communications services, which are essential to the functioning of public institutions and administrations;

– community-controlled communications services (infrastructure, waterways, ports, airports, hospitals and public education).

The concept of public service is guided by the "subjective" approach to the Council of State's jurisprudence. It covers the willingness to ensure, whatever the circumstances, uninterrupted service offered to all citizens without exception and to compensate for the shortcomings of market offers (without proposing competition). On the other hand, the French definition of universal service has been considered broader than the European vision [NGU 00].

4.2.2.1. *The concept of universal service*

Universal service is based on three main principles:

– availability: the level of service must be identical for all users in all places and at all times;

– affordability: the price of the service must not be prohibitive;

– accessibility: subscribers must be treated equally in terms of service, price and quality of service, in all places and at all times.

A revision of the rules of the aims of public service and universal service may be considered in order to take account of changes to the market and to technologies or of the needs identified for a sustainable practice (social service). The French territory is vast and the population is concentrated around certain centers, leaving large areas that are expensive to equip. On the other hand, using digital services requires a learning process and continuous support, which elderly or isolated people cannot access. "Digital illiteracy", whether partial or total, reduces the scope of public investments in favor of new technologies. In rural areas, the gradual closure of post offices and the difficulty of setting up Internet cafes inspire regret at the loss of the former cooperation between postal and telecommunications services. Agreements on international trade, however, underline the need to take the specific characteristics of each country into account in applying these principles. As a result, the regulations on the concept of universal service and public service in communications must be adapted to societal factors in order to take into consideration the changes brought about by technology, customer behavior and the need to support investment and national innovation. After several decades of privatization, the balance sheets of networked companies, which have taken up the liberal position, are far from convincing [JEA 15, VAN 15].

In the United Kingdom, the British regulatory authority Ofcom has established a high-speed universal service obligation (USO) to enable citizens to have a 10 Mbps connection, wherever they live. In 2015, three and a half million households and businesses in the United Kingdom have already had access to data rates of 24 Mbps or more. The United Kingdom intends to provide 95% of households with very high speed access by the end of 2017, despite the existence of numerous technical difficulties caused by geographic and demographic circumstances.

At the European level, and for those who are not legal experts, the regulations that have been developed to "protect the interests and rights of end users" seem incomplete and quite uncoordinated from one country to another. It seems that initially worthy sentiments on the issue have been reflected in the clarity and simplicity of the regulations. Certainly, technical developments occur rapidly and their spread is unforeseeable. Within Europe, in 2016, universal service is understood quite differently in each of the 28 countries and it is likely that the gradual demotion of this aim is linked to the decline of the required funding [BOU 16].

In the United States, they are currently moving toward replacing some payphones with Wi-Fi hotspots with high speeds of 1 Gbps. More than 7,500 hotspots with ranges of 45 m will be installed in New York, and will be equipped with screens displaying advertising and city information as well as users to call the emergency or municipal services. These public booths, financed by advertising (some of which will come from Google), will allow portable devices and tablets to be charged. Some booths will even be equipped with a tablet. This development, linking network operators to advertising agencies and OTT services, could be a means of developing a broader implementation of universal service within this framework.

Public service delegation is used for the construction and management of communications infrastructure, for example to connect an island to a continent, or to create a fiber-optic network in an urban or industrial area.

4.2.3. *People with disabilities and under-age children*

Public facilities must be accessible to people with disabilities. Communications services, in particular, must follow this arrangement. Special terminals have been designed for people who are blind, deaf or hard of hearing. Communications services designed for deaf people or those with speech impairments are also operational among some network operators.

In the United States, Web companies must respect the Children's Online Privacy Protection Act (COPPA), which forbids them from collecting personal information on young people under 13 years of age without the permission of their parents. In theory, the majority of social networks do not allow users to be under 13 years old. In Europe, this rule is not reflected in national regulations. In France, the prior consent of parents is only required in cases where a website collects a photograph of a minor, or uses his or her personal data for advertising purposes. At the end of 2015, the European Council suggested that this age limit should be raised to 16 and to leave it to each member state to choose the age of consent, between 13 and 16 years of age. A European document on data protection in Europe should be adopted by mid-2016, and then written into the different laws of member states.

The ISO/CEI Guide 71:2014 provides suggestions for standardization on how to handle demands and recommendations relating to accessibility, which, directly or indirectly, target systems used in computing and communications. The ITU and the ETSI have also drafted standards on this topic. While one group of people with disabilities may benefit from new facilities in communications practices, there is presently still much to be done to make these upgrades more widely available.

The International Bureau for Children's Rights (IBCR), an international non-governmental organization (NGO) located in Canada responsible for protecting the rights of children, is calling for public authorities to exercise greater vigilance with regard to the Internet. It has mentioned that, "the fast expansion of the Internet has not only made it easier to buy and sell sexual and pornographic services, but it has also created an environment favorable to sexual interactions. The Internet has become a place where children explore their burgeoning sexuality. Furthermore, the Internet has broadened the scope for sexual exploitation of children by facilitating their direct and anonymous contact with adult predators".

With regard to the content of public services available on the Internet, regulatory bodies (including the CNIL in France) are demanding that specialized websites protect the personal data of children aged between 6 and 12, as well as of teenagers. The websites accessed by young people variously include social networks, games, educational sites, academic support sites, TV channels and news sites. Often, the websites visited are not suitable for the age of the children. On the other hand, the vast majority of websites collect personal data (IP address, location, email address, geographic coordinates), and a minority of these sites provide the option of deleting personal addresses. The majority of these websites are not suggesting any monitoring measures for young people, or for parents to ask for their agreement. The French body CNIL has underlined the shortcomings of monitoring and parental control in this area.

The WHO wants use of the suffix ".health" in electronic addressing to be associated with the establishment of justifiable codes of conduct by effective and applicable accountability mechanisms under the relevant government authority. These codes of conduct should correspond to all concerns related to healthcare on the Internet. They deal with the quality and reliability of sources of information on health; trust in products, services and practices; and combating illegal activities and deceptive, malicious and underhand practices.

The Swiss "Health On the Net" (HON) foundation guarantees the annual certification of health-based websites, founded on respect for deontological principles, which include the reliability and credibility of medical and health-based information. The French National Authority for Health (NAH – *Haute Autorité de Santé*) is planning to suggest "methodological reference points to editors of health-based Internet sites and Internet users for access to reliable information". However, the NAH has not negotiated any convention with the HON foundation since 2013. This project should therefore be constructed in tandem with healthcare professionals, public authorities, users and editors. Nonetheless, in 2014, close to 882 French-language Internet sites (out of over 3,000) were still receiving the benefits of "HON Code" certification (www.hon.ch/).

The lack of Internet governance by a responsible authority currently leaves Internet users without information on serious aspects of the sites they visit and on the possibility of all directly accessible content being available (see section 5.1.1.5).

4.2.4. *Security of transactions*

Since Edward Snowden's revelations about the PRISM scandal in 2013 (section 4.6) [BAT 14], everyone knows that the Internet is a network designed for easy communication, but that it provides no security. Every year, the press reveals dozens of significant breaches of personal or banking databases, due to the permeability of the network and the lack of precautions taken by managers and Internet users. Currently, the Internet remains a network on which security is not guaranteed and it consequently falls to users to take responsibility, pending new technical measures or the availability of more effective individual protection.

The NSA maintains that its actions are entirely within the law, with regard to the wiretapping and surveillance of the Internet and mobile networks. All countries are victims of computer espionage and the majority of American businesses are victims of cyberattacks. Most of these crimes are often little known among the public. Big Data companies, which stock and manage large volumes of data, are the most exposed to cyberattacks. Tomorrow, the Internet of Things could become the target of organized cybercriminals. It belongs to all businesses to estimate the risk precisely and protect themselves against this threat.

Protections against wiretapping are faultless if they are comprehensively carried out, but, in such cases, the content of the message can only be intercepted through possession of the data encryption key. If it is a case of terrorists preparing a guerilla attack, the authorities trained to monitor messages will not be able to decode them. All that they could obtain would be a scrambled message. Policy makers must continue to question cybersecurity experts to find a response to this situation.

In France, since the end of 2015, the legal branch of the national police has been receiving the assistance of a unit dedicated to combating all forms of Internet-related criminality. This unit is concerned with combating digital crime, meaning that it focuses on terrorism, Internet fraud, computer piracy and bank card piracy. This unit carries out research on the conventional Web or on the "Darkweb". It also handles many issues linked to the Blockchain with regard to smartphone applications or messages hidden in online video games.

4.2.5. *Internet neutrality*

4.2.5.1. *Definition of neutrality*

Through the phrase "network neutrality", regulatory authorities are expressing a principle guaranteeing the equal treatment of all data rates from the moment they enter the network. This equal treatment must be understood both technically (no discrimination with regard to the source, destination or content of the information transmitted) and economically (the data rate must be accessible at the same price for all content providers). "Net neutrality" is an important concept, as it is linked to the quality of connections (VoIP, data, video), with good quality only guaranteed at a high price.

4.2.5.2. *Technical aspects of "neutrality"*

The desire for "non-discrimination", which has been expressed since the beginning of the Internet, is in fact a theory and, for around 30 years, this contradiction has been raging without examination of the technical realities. Naturally, the provider of communications channels in IP protocol would like to have the largest possible number of satisfied customers. In order to attain this, their varied requirements must be taken into account. There is no question for the provider with regard to favoring one or other of its friends or partners; on the contrary, it must ensure that the massive flows to a broadcaster of television programs does not obstruct the security service provider seeking to avoid delaying or cancelling the transmission of emergency alarms due to temporary congestion on the network. The network operator must therefore organize the traffic in separate channels to avoid any possible difficulties, which naturally leads to different cost prices and therefore to an infringement of the neutrality principle. In fact, everything could be confidently carried out if Net neutrality was understood in the sense of "non-discrimination in the event of equal traffic" (see also section 2.2.3.2).

4.2.5.3. *Initial organization*

After lengthy discussion of this topic, American regulations have come to accept specific treatment for the broadcasting of videos on the Internet by cable operators, in return for a financial contribution and thereby organizing a two-speed Internet. In Europe, the situation is different with regard to broadcasting video, probably due to the linguistic division of the market, and in October 2015, the European Parliament voted on a text protecting the quality of the high-speed transmission services planned for teleconferencing and telemedicine. This regulatory agreement, which was adopted by each of the 28 European Union member States, leaves operators the possibility of slowing down the data rate to avoid network congestion, provided that all of the management measures are "transparent, non-discriminatory and proportionate", and do not last any longer than necessary. This compromise also safeguards the quality of services.

4.2.5.4. *Points still to be clarified*

By maintaining the "ambiguity" of the concept of "specialized services" likely to benefit from priority network access, network operators will have an argument for bypassing Net neutrality.

The impact of the new provisions taken in developing the connected Internet of Things is yet to be clarified. Cases of "vital" applications with low latency times in mobile telephony do not appear to have been taken into consideration. The BEREC, which brings together European telecom regulators (section 3.2.2), has been tasked with developing "general guidelines" on Net neutrality by the end of 2016 to avoid differing interpretations of the text from the European Parliament.

To this effect, one law, currently being prepared in the Parliament, is intended to clarify the French position on European regulations. Among the major principles proposed, this act should highlight some of the key elements, such as "the freedom of innovation". This act must also clarify the French concept of the "principle of Net neutrality", understood as the absence of operator discrimination regarding the content on its network and the "loyalty" of computing platforms connected to the Internet.

This indicates that common rules regarding the Internet should soon be available at the European level, contributing to a single market without fragmentation. It should be noted that, with the two aforementioned principles, it would be difficult for the network operator to correctly size its links and routers if the actors in charge of applications have not previously defined the features of their traffic, with the Internet network not carrying out any dynamic data rate allocation on demand or by reservation.

4.2.5.5. *Tactile Internet and "Net neutrality"*

On the other hand, it appears that, in certain areas, the debate on Net neutrality is somewhat outdated, with several applications requiring use of an Internet network in almost real time (to the nearest millisecond), which currently only the "Tactile Internet" is able to provide at distances under 150 km, pending the arrival of a quantum Internet.

4.2.6. *Respect for personal data*

In each of their actions, citizens leave traces of their activity and these traces can be stored by the employees of popular bodies (post office, insurance, shopping). By the same token, Internet users leave on the network traces of their addresses, personal data and visiting patterns for different sites. The permeability and automatic memory of the Internet means that it is easy to profile each Internet user and use these profiles for marketing operations or market research.

The detailed information of Internet users can be harnessed for competitive purposes by marketing professionals, insurance agents or the managers of GAFA companies (Google, Apple, Facebook, Amazon) and social networks. The door is open to innovative and competitive action in these areas. As it happens, Internet users are no more the owners of their personal data than the banks they trust or their mutual health insurance plan. The paradox lies in the fact that, conversely, Internet users, despite everything, have the advantage of being able to use their personal data to enhance their community's wealth of healthcare information. Their medical history is useful to others like themselves, just as their consumption habits assist in the organization of a coordinated regional supply system. While their shopping habits are known to providers, Internet users are in some way protected against the actions of impersonators who seek to use these factors to steal from users. Through the information stored, banks can uncover deceptions and block any attempted fraud. In reality, the discussion lies between the potential misdeeds of an artificial "Big Brother" and the paternalism of a benevolent "Godfather".

The priority status afforded to client-oriented services considered to be "profitable" is a constant feature in free trade. The novelty brought by the Internet is linked to the dynamism and power of the computer tools used, which have revolutionized habits and which have significant storage and information management capabilities. It is therefore necessary for Internet users to consider both the positive and negative aspects of their personal data being used by sites that they frequently visit.

The discussion has already continued into 2016 and all evidence points toward past abuses being corrected by more restrictive future legislation. The CNIL has recently given Facebook a 3-month deadline to stop storing the data of people who do not have an account on that social network. Several regulators in the European Union have begun inquiries on Facebook's actions toward protecting the privacy of individuals.

Within the framework of Internet neutrality, the Indian communications regulatory authority (TRAI) has forbidden mobile networks from offering free differentiated access to the social network Facebook. There is no reason, according to the TRAI, to offer data services for different prices on the same network.

Moreover, Facebook, due to its paternalistic attitudes, has suffered setbacks in other countries. It is intermittently blocked in several countries in Asia and the Middle East because its content is deemed discriminatory with regard to religion. The security of accounts and personal data has been compromised. On the other hand, Facebook is contentious as it can cause disturbances due to people using it during working hours and frequent instances of rudeness from people witnessing irresponsible behavior. Finally, like most OTT, Facebook carries out undisguised tax evasion practices.

4.3. Remote payments

Within communications applications, this sector of activity has suddenly grown to an enormous scale, not because of the sums of money involved, but due to the speed and security of money transfers that are now possible, even on the move or in developing countries.

4.3.1. *Currency and remote payment*

4.3.1.1. *The triple role of currency*

Since the days of Aristotle, currency has been defined by the triple functions; it is guaranteed by each sovereign State. It serves as a unit of account, a store of value and a medium of exchange. Currency allows the payment of goods and services as a function of being legal tender, which is the result of the total of social, economic and political conventions based on mutual trust between partners. Economic theorists, including Adam Smith, Ricardo and J.-B. Say to name but a few, reduced the role of currency to that of a simple intermediary agent in economic exchanges. More recently, J. Keynes underlined its power in the economy. Currency is a store of value and a social commodity that, through its circulation, galvanizes the employment level and growth of a country.

Different forms of currency can be used at the same time in the same territory. Currency areas are jointly defined by government authorities. The International Monetary System (IMS) manages exchanges of national currencies between agreed parties, which today are no longer linked to a gold or monetary standard (Jamaica Agreements, 1976), but to bilateral floating values determined by the IMF.

In support of regulatory texts, national authorities can be authorized to create additional or replacement currencies to be used for a limited period (vouchers, meal tickets, treasury bills, future values, promissory notes, stamps and transfer forms from the post office or for stamp collectors, telephone tokens, plastic cards for phone calls of a fixed duration). A currency's credibility relies on a general acceptance, the threshold for which is connected to the aims of the community that uses the currency. Banks, the post office, communications network operators and businesses create currencies. There is a strong trend to create a dematerialized currency that is able to avoid tax authorities, control and various taxation systems, which are always considered excessive.

To illustrate the importance of this market, PayPal's online payment activities have connected 175 million consumers with 13 million merchants. Because of the development of mobile phone technology, this customer–retailer relationship can be enhanced further.

Monetary circulation is created by the credits that support the borrower's actions. However, as far as currency creation is concerned, banks cannot pass a certain threshold, with the risky assets not exceeding a percentage determined by the bank's equity [COU 14, PLI 13].

4.3.1.2. *Special drawing rights*

The IMF decided that from October 1, 2016, international exchanges, after having used the reference points of dominant countries (Pound Sterling, Deutschmark, Yen, Swiss Franc, CFA Franc), will be based on a "reference basket of currencies", defining the special drawing rights formed as follows: 42% in dollars, 31% in Euros, 11% in Chinese Yuan (or Renminbi), 8% in Japanese Yen and 8% in Pound Sterling. All countries in the world use this for the international trade of goods and services, with this reference point having the advantage of being a relatively stable temporary arrangement.

4.3.2. *Electronic trade*

4.3.2.1. *Definition of electronic trade*

Electronic trade (or "e-commerce") uses the medium of communications networks. Due to its specific challenges, it has a new approach to customers. Different aspects of transactions, including collecting the taxes payable in place of delivery, monitoring transactions to protect customers and respecting national regulations, must be clarified further by national regulatory bodies.

4.3.2.2. *Securing online transactions*

Several solutions have gradually been developed to secure online transactions. Among these, the "tokenization" technique in remote payment solutions opens the way to possibilities that can be easily implemented and are capable of offering customers a simple, fast and secure ergonomic purchasing procedure.

The tokenization process is carried out by the card reader that, through the bank card's number and expiration date, determines three control values that are transmitted to the Internet server via the network: the three figure security code on the back of the card (card verification value (CVV)), the expiration date and a number, known as the "token". This "token" serves to make transactions secure and provides protection against fraudulent use of payment cards. As of 2016, new bank cards have a CVV that is randomly selected and can be renewed every 20 min. This new CVV is visible in a small window located on the back of the card (dynamic cipher).

Recently, technologies for dialog between users and Internet servers have been added to Web browsers. The Ajax (asynchronous Javascript and XML) method allows rich Internet applications to be installed, offering a greater level of manageability and ease (so-called "Web 2.0" applications). Internet browsers have been made compatible because of the availability of JavaScript code, which can be integrated within Web pages and executable on the client's terminal. JavaScript, used in the Ajax method, is used to modify the content of Web pages to anonymize the process within commercial databases. In this way, transmitting the bank card number of the payment services provider (PSP), performed using Ajax, does not pass through the merchant's server. In this way, the card's security code, as well as the sensitive data linked to authorizing transactions, cannot be recorded by merchants.

4.3.2.3. *Criticism of the standards used*

A binding regulation, administered by the Payment Card Industry (PCI) Security Standards Council, has been created in order to reduce the fraudulent use of payment instruments. The PCI DSS (Data Security Standard) standard was established by payment card providers and version 3.1 of it was published in April 2015. It specifies 12 conditions of compliance, divided into six groups called "control objectives", which are a substantial burden (there are more than 300 "good practices", with equipment auditing and implementation procedures). Setting up a collaborative network for monitoring fraud allows the repetition of activity to be monitored and guarantees a good level of protection.

Tokens are stored within electronic wallets and smartphones. They are at the center of numerous integrated solutions in the purchasing process. Interbank networks could certainly standardize this token method, with or without smartphones.

4.3.2.4. *The international framework*

E-commerce is progressing slowly. It appears to be difficult to connect identity verification to user-friendliness in light of the security that has become a major concern. The exchange rate between currencies and the interbank commission to be paid for the requested transaction is also slowing down adoption of this procedure.

The European Commission would like to quickly provide Europe with a digital single market (DSM), constructed to assist European businesses in preparation for their own growth and the growth of the labor market. In 2015, 7% of European SMEs sold abroad online and 15% of European consumers bought online from outside their own country. These low percentages are explained by delivery costs, existing regulations and the different VAT rates in force in the 28 member States. Digital exchanges of cultural goods are restricted by a complicated system for dealing with copyright that is related to territoriality. Lastly, according to European Commission research, online trade has been slowed down by the low density of wired and radio-based broadband connections in Europe and the fact that Europe has fallen behind in Cloud technology and in managing massive amounts of data. Drawn in by the supply and frequency of purchasing, the expenditure of French online shoppers continues to grow and should rise to 70 billion euros in 2016.

The practice for sharing the responsibility for actions undertaken in this area is still poorly defined and the DG Connect Commission seems to have forgotten to appoint someone capable of bringing the messy written regulation into line on the technical and legal fronts.

In February 2016, the ITU-T prepared a draft standard on MFS (mobile financial services) (see Appendix 11).

4.3.2.5. *Electronic wallets and mobile payment*

It was banking networks that had the initiative to launch a digital wallet. However, mobile devices appeared perfectly designed to serve as focal points for the service known as "mobile banking". In this way, the "Orange Money" service, launched in Africa and the Middle East in 2008, has acquired, in the 13 countries where it was established in 2015, some 14 million customers, a number that should rise to 30 million in 2018. "Orange Money" assists in the transfer of 500 million euros every month and also guarantees the loyalty of its customers. In 2014, this service generated around 50 million euros in revenue. The operating method requires that a set of rules on setting up "tokens" be followed. According to countries, banks and network operators are either competing or collaborating in this service offer, with or without recourse to the Internet. In Kenya, in 2013 the activity of "Mobile Money" represented close to 43% of the country's GDP. In India, the three biggest mobile network operators opened bank accounts for a total of 400 million mobile subscribers. In 2015, Spanish, Polish and Turkish banks optimized their customer pathways with a videoconferencing service. The number of users of mobile banking services could double between now and 2019, reaching 1.8 billion people around the world (KPMG) and, in 2017, mobile phone applications could produce a global sales revenue of 77 billion dollars, with each mobile user able to send personalized information to more than 100 services each day (Gartner Inc.).

4.3.2.6. *The case of the stock market*

Stock market transfer orders made over the Internet are often associated with software such as Acorns, Digit.co, Level Money and Mint, which allow users to quickly ascertain the general overview of transactions made at the end of a connection session. However, the ability of stock market applications to do this presents a risk to customers due to the potential vulnerability of connections established without contractual security (particularly where movement is involved), and, due to its success with the public, it creates a great changeability in pricing, with the speed of execution and the simultaneousness of orders able to cause panic among market observers.

Accused of systematic manipulation for nefarious purposes, the stock market's computing automatons (automatons for high-speed trading, also called "high-frequency trading") are seen as the "barbarians" of Wall Street and act shamelessly. This technology is accused of assisting in dubious market operations, price manipulation, "shadow markets" and the sharp fall of Wall Street prices, which sparks plenty of controversies [LEW 14].

4.3.2.7. *Virtualized transfers*

Alternative financing arrangements, which are made using prepaid bank cards, and the practice of microfinancing, or "crowdfunding", allow tax and intelligence services to be evaded. On the other hand, the Nickel account, arranged by tobacconists, goes through a triple verification process for the client's identity, and the movements of each account are monitored in real time to track suspicious behavior.

4.3.3. *GAFA and online sales*

Online sales greatly interest the GAFA companies, innovative Internet players with considerable economic and political power, which have memorized the personal data of a large number of Internet users and acquired dominant positions in ICT markets and on the Internet. The GAFA companies help one another to promote their services or related goods created by their subsidiaries or associated companies.

Facebook (which alone is used by a billion users per day) and Google take up 25% of the time spent by Internet users on mobile applications. For its part, Facebook, which had a turnover of 12.5 billion dollars in 2014 (of which 78% was mobile advertising revenue with 2.5 million advertisers), is testing a new type of page that would allow products to be purchased. It has already tested a buy button in 2014, and it officially launched money transfers on its messaging application Messenger at the start of 2015.

Google has set up a feature allowing its users to purchase products directly through the results of searches made on mobile devices, without having to go to another merchant. Google's economic model uses sponsored links paid for by businesses and on which commission is paid. This system, which works on desktop computers, mobile devices and tablets, is always highly successful.

4.3.4. *Contactless payment*

The system of remote payment with an NFC (near-field communication) payment card uses an NFC radio system, which is brought into play between the buyer's card and the merchant's terminal. One simple application, freely downloadable from the Internet, allows the confidential information contained on a bank card equipped with NFC technology to be accessed via an MSC-compatible smartphone. By placing the mobile telephone on the card, the card number and expiration date, and therefore the two pieces of information required for payment, are readable. Indeed, the MAC (message authentication code), intended to prove that the data come from a trusted third party, is poorly positioned in IT terms.

Moreover, the merchant's terminal sends its user name through information exchanges with the body handling the payments. This connection is encrypted, but it can also be read on each printed receipt. Mistakenly, the password of the encrypted link is generally the same for all of the merchants connected to the same payment management organization. A hacker could uncover it using special software in a matter of seconds.

Use of payments without NFC contact can only move forward by increasing the amount of reading equipment in businesses and by strengthening security systems.

4.3.5. *FinTech*

New players have recently appeared in the digital sector – "FinTech" organizations. These are corporate law firms that combine their profession with basic expertise in cryptographic software, antifraud systems and bank card management. At the same time, they encompass four branches of regulation relating to payments, banking law, asset management and insurance, considered at the national, European and international levels.

Because of the Basel III agreements, which came into force in December 2010, FinTech organizations have been able to develop in Europe, as they are not subject to the same cumbersome regulations as banks. On some aspects, traditional banks, FinTech organizations, network operators and the GAFA companies find themselves in an increasingly international remote payment environment, either competing or obliged to accept a partnership. As both the competitors and allies of banks, FinTech organizations use the resources of banks' client bases.

FinTech organizations are developing computer applications (API) that only require a small volume of data and allow fast access to massive amounts of information through the Cloud technology. It will likely be banks and FinTech organizations that, together, after examining the available solutions, will select the most suitable system for the biometric authentication required for secure bank transactions. Alternatively, they will at least determine the best usage of the Blockchain.

The aim is to reduce costs and timeframes, while increasing security and reducing losses in order to attract customers. A bank's founding principle is to gather funds to finance its activities. However, the growth of players in this sector is causing instability in customer capital. The key concern of banks is currently to create customer loyalty to avoid losing funds.

The European Parliament is aiming to encourage market access among new players, in order to create competition, which will be beneficial to consumers. In the second version of the Directive on Payment Services (DPS2) from October 2015, the existence of FinTech organizations was validated, including an expert assessment consisting of collecting and collating the banking data of their customers in several financial establishments and using them in applications. In 2014, according to Venture Scanner, over a thousand FinTech organizations were active globally, four times more than in 2013.

4.4. "P2P" exchanges

The P2P communications system must not be confused with either the term "point-to-point connection" or the "point-to-point" network protocol (PPP). Here it concerns a direct exchange of files between computers equipped with the same software.

4.4.1. *P2P, Blockchain and Bitcoin*

P2P communication is used to share files in P2P (BitTorrent, for music or video), for grid computing or for communication (streaming media, Skype and TeamViewer, for example). Downloading a file on a P2P network does not allow either the file's creator or the users connected to the network to be identified, which is a security threat. The sharing of P2P files is replicated on a large number of hubs, which reduces the burden on demand and eases network traffic. This system makes censorship or hacker attacks more difficult. On the other hand, it is impossible to measure the volume of traffic (section 2.7.2.2).

It is said that, in 2008, Satoshi Nakamoto, a Japanese computer specialist, invented a piece of software (known as Blockchain) that is able to make exchanging virtual currency in Bitcoin faster and less expensive without the assistance of central or commercial banks. The Blockchain software is a secure software program with P2P encryption, which uses an information hash feature to encode information, with the elements being dated and stored in the memory of several sites simultaneously, thereby guaranteeing the authenticity of the message. The universal database records transactions and grows in a linear fashion in chunks known as "blocks", forming the "Blockchain". The Blockchain can also operate with encryption technologies. This allows the full traceability of transactions and prevents counterfeiting. All transactions on Blockchain are public and the user is not obliged to disclose his or her identity completely. The first application deals with transferring funds in units called "Bitcoin" between partners on the P2P network. Bitcoin has become at once a currency, a technology, a network and an open accounts book.

4.4.2. *Alternative cryptocurrencies*

In the past, thousands of instruments for monetary exchange were developed. Since 1980, several projects aiming to transform the world of economics have supported the creation of local, social and complementary currencies, known by the acronym "LETS" for "Local Exchange Trading System". Some are based on the idea of the time used (the Canadian "accorderie"), others on solidarity and still others on promoting local development (of which "LETS" is an example). The roughly 200 LETS, which have been set up globally, still cannot be converted and can only be used in a 50 km radius [SER 99, KEN 08].

There are still several hundred digital currencies that claim to be usable as methods for virtual exchange without benefiting from State liability. "These virtual currencies circulate on P2P networks through cryptographic algorithms. Such currencies include Reddcoin, Litecoin,

NXT, the Hayek (based on the value of a gram of gold) and Bitcoin. These trading tools can be classified according to their uses: socially oriented currencies, currencies intended for infrastructure, alternative currencies and currencies for targeted purchases" [FIE 14].

4.4.3. *Other Blockchain applications*

"Blockchains" allow financial transactions to be tracked through a decentralized IT system: incredibly powerful computers, known as miners, verify that transactions are real and that no imposter has slipped through. Transactions are confirmed by blocks and then added to the record, forming a chain or blocks, hence the name "Blockchains". Every computer connected to the network hosts a copy of all of the exchanges made. Some user volunteers run a computer program to update the record and detect anomalies. The anonymity of users is protected.

Since 2014, the development of the North American Ethereum project, similar to the Bitcoin project, seems to indicate that it is possible to simplify all slow and expensive authentication procedures, such as those of insurance companies, notaries, Wall Street, online film distribution companies such as Netflix and Hulu, gaming platforms, messaging services such as Twitter, currency exchanges and so on. The total amount of Bitcoin transactions, supported by the current Blockchain software, is around 3.4 billion dollars.

Originally designed for Bitcoin, the Blockchain IT network appears destined for some development. The P2P principle can, according to the promoters of collaborative economics, create value, namely common, tangible or intangible property, without being compelled to ask for permission [BAU 15]. It could transform global finance by removing the "trusted third parties" who currently verify transactions (notaries, clearing houses and so on). Several possibilities are currently being assessed for activities that could connect order executions to the safeguarding of securities[1].

1 https://block chain.info.fr/.

Several study groups (Agora, BitCongress and Swarm) are researching the possible use of Blockchain for online voting. Blockchain technology can also be used to keep track of the different owners of diamonds, boats, watches or works of art, any object which must be registered for life in order to be insured. This project is also an attempt to try and eradicate fraud and theft. Land registry management, patent administration, accounting, commercial "couponing", music and journalism could use "Blockchain" systems.

Presently, Bitcoin only allows seven transactions per second and 10 min must elapse before beginning a new transaction. A new Bitcoin XT differentiates itself from the original version though a larger memory to store the traces of transactions, manage a greater number of transactions and increase network speed. However, the medium size of blocks has continued to increase and is currently approaching 700 kB. On the other side, networks are becoming unstable and endangering trade. Yet, network reliability was one of the main arguments in favor of Bitcoin. Without a significant agreement in 2016, this virtual currency risks disappearing in favor of other competing digital currencies (Ripple, Litecoin and so on).

Regulatory bodies are trying to understand technology and better grasp how it functions to guarantee the protection of citizens by creating the most suitable playing field. P2P technology is legal and increasingly used in various business models (European Commission, https://ec.europa.eu/digital-agenda/en/glossary). However, a warning on the usage of these currencies was given by the French Ministry of Finance in 2014[2].

The main problem is that virtual currencies such as Bitcoin are not issued or guaranteed by a suitably powerful authority, or by a government, for example. The disadvantages of Bitcoin – its instability and the appearance of unexpected technical problems – explain why it has not achieved the level of acceptance that venture capital companies expected.

2 http://www.economie.gouv.fr/files/rapport_ monnaiesvirtuelles_web.pdf.

4.4.4. *Banks and P2P*

4.4.4.1. *Features*

Blockchain is quickly becoming not only the most controversial, but also the most promising technology in the world of financial services, where it sparks as much confusion as curiosity. The interest of banking institutions and FinTech organizations in Blockchain lies in the potential to reduce fees and transfer times, as well the number of actors in financial operations.

4.4.4.2. *Financial markets*

Blockchain technology seems very promising for the financial markets (NASDAQ, for example), as it is more efficient and both more transparent and more secure, which is essential. It is an IT breakthrough distributed on the basis of hash functions. Blockchain technology is able to bypass financial intermediaries and avoid fiscal controls (and the potential financial transactions tax (FTT), or Tobin tax, currently being discussed by the G20).

The banks Santander and Barclays are looking for their own means to improve their services through distributed and accessible servers in trusted institutions. This money transfer technology would allow infrastructure costs to be reduced by 20 billion dollars per year in total, if the sector accepts a thorough examination of its current technology.

IBM would like to establish a payment system based on Blockchain and on a more efficient and less expensive infrastructure. Partnerships could unite the Fed, IBM and Apple, leading to an open and less expensive payment system that could give a competitive advantage to American players in the field of payments.

Bank of America has filed a patent for a system and method for wire transfers using cryptocurrency on the basis of several systems (including OKCoin, BitStamp and BTCChina). The NASDAQ has implemented a Blockchain to facilitate transfers and sales of shares on its private market. Some large banking institutions, such as Barclays, Visa Europe and the Royal Bank of Canada (RBC), are also interested in Blockchain. In 2015, around 20 companies were guaranteeing the international transfer of funds in Bitcoin, particularly in the United States and Asia.

At the end of 2015, the Chinese Banking Regulatory Commission reduced the activities of some of its online banking platforms. From now on, P2P sites should provide specific information on the quality of lenders, the outstanding amounts of borrowers and the risks of investment projects. China does not wish to exhaust a source of investment, which is so useful to the economy. During 2015, almost 57 billion euros were loaned to Chinese SMEs through P2P sites.

4.4.4.3. *Hidden loans*

The P2P platform financial service aims to democratize banking practices beyond conventional channels by simplifying procedures and sharing the risks through a network or partners, which would separate the needs of individuals from those of institutions. The sums of money involved are relatively low:

– a Lending Club is a company for participatory finance between individuals registered with the Securities and Exchange Commission (SEC);

– Funding Circle is a P2P loan platform that allows loans to be directly allocated to small and medium-sized businesses over 5 years.

Crowd Funding – due to the high level of banks' management fees, alternative loans in P2P seem to have a promising future around the world. Since 2005, market loan platforms have offered an alternative to banking systems and traditional payments. Around 20 French platforms were operational in 2015 for local projects. The amount of loans is close to the amount of microloans (4,500 euros on average, for a total of 140 million in comparison with 4,400 billion in savings in France in 2015). The regulations should soon be clarified, as well as the related taxation.

GAFA – the creation of financial platforms for shared financial operations (crowdfunding) in Blockchain is currently being researched by the GAFA companies. Facebook has obtained accreditation to offer

P2P payment services in the United States by bank card through its own P2P messaging service.

4.4.4.4. *Potential risks*

The risk inherent in the process is connected to its originality, as anyone can join the group and an organization with sufficiently powerful IT calculation abilities could modify its software. As there is no supervisory authority for this sort of activity, it is possible that Bitcoin will remain restricted to minor activities, with the risk of the proliferation of versions of Blockchain software. Virtual currency transfers could be used by drug traffickers, or hackers of online purchases in foreign currencies who finance terrorism or seek to evade the watchfulness of the authorities or the chain of intermediaries. On the other hand, Blockchain technology is likely to infringe the financial sovereignty of States.

Like the Internet, Bitcoin will develop further. However, mass adoption of Blockchain seems further than ever from being achieved due to the numerous psychological, cultural and technological obstacles that have been identified.

4.5. Remote computing

The click-through rate (CTR) of a website reflects the activity of Internet users. In this way, the histogram in Figure 4.2 shows the hourly visitation of a site located in France. It shows that, in reality, over half of customers (55.6 %) are located in traditionally French-speaking time zones, which is confirmed by the majority of the Internet users' IP addresses. The Internet is indeed a global "village" where information is exchanged at all hours of the day and night. In this example, 44.4% of requests come from countries very far away from France. It transpires that this site has an audience that very happily matches its purpose. Conversely, when dealing with purely local activity, it is worth trying to ascertain whether the installation of an Internet site is indeed justified or if it would be preferable to turn toward a mobile application focused on the audience, which would be simpler to implement and more efficient.

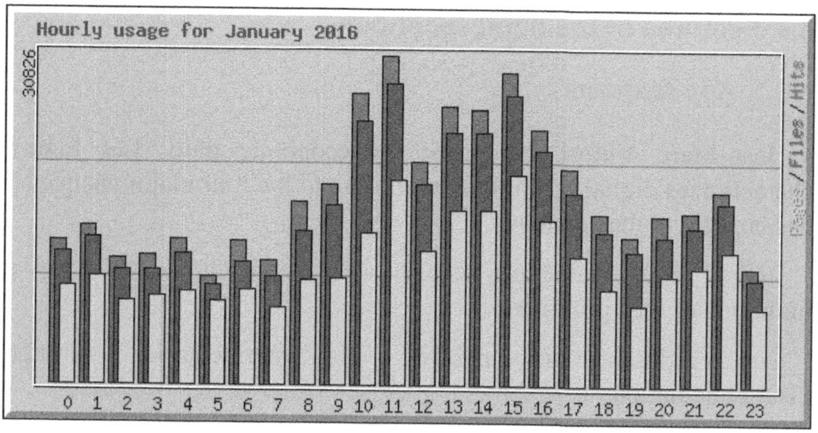

Figure 4.2. *Hourly click rate on a French language website*

ICT development is currently dealing with the production of all things and the management of all services. It is modifying traditional approaches to innovation, organization and production. It is bringing about new activities, breaking with existing activities and thereby giving new life to Schumpeter's concept of "creative disruption". The use of IT in the business sector has been integral to moving the concentration of points of sale and creating large retail spaces. Investment dedicated to new technologies, and therefore to software, should not be forgotten as it produces a significant reduction in working time, energy consumption to the customer's benefit and the dissemination of knowledge.

Software is eating the world; so said Marc Lowell Andreessen, founder of Netscape, in 2011. This assessment simply continues to develop as the world becomes more digitized. According to Afdel, in 2014 the 100 leading French players in software saw their sales revenue jump 10% to 5.6 billion Euros, as opposed to just 2% the previous year. Ten years after its launch, SaaS (software considered to be a service) has undergone significant development. The banking sector and insurance policies will finally benefit from applications that are more secure than in the past.

4.6. Features of the digital economy

4.6.1. *Key features*

Jean-Marc Vittori, columnist for economic daily Les Echos, characterizes digital technology according to the four major changes it has brought to the economy:

– first, digital technology leads to a very wide choice offered by a multitude of web providers;

– it reduces the number of intermediaries through disintermediation;

– third breakage, some links in the participant's chain are voluntary;

– finally, if a digital service has a development cost, its dissemination is very inexpensive.

It is possible to add other considerations to these features attributed to digital technologies. Digital technology allows information to be distributed among a large number of players more quickly and widely. By the same token, digital technology sets off a large number of reactions and subsequent positive actions. When innovations are increasing in all areas and at a faster pace than before, this acceleration is due to a set of digital means that are now accessible to all. This spread of information is carried out at a very low cost and the number of barriers restricting access to information is also increasingly low.

There is no longer any public report available on the transition of the French publishing sector to digital technology. Each media group will undoubtedly carry out its own internal research on this development as digital technology allows economies of scale to be made, as well as savings in information storage and in production and labor costs. However, the transition from paper editions to entirely digital editions poses a large problem for daily and monthly newspapers with regard to the adaptation of their readers, be they connected through a fixed-line or mobile communication system. The website of the 2016 French press and media watchdog (*l'Observatoire de la Presse et des Médias*[3] notes the very low circulation rates of digital newspapers (under 2%). For now, in 2016, so-called "digital" editions are still largely considered a specific service reserved for subscribers who are willing to be charged. Nonetheless, everyone can agree that progress in this area is slow but steady.

4.6.2. *Preferred sectors*

Nonetheless, the digital economy has only introduced its innovations in a liberal environment for specific activities and for a limited clientele. However, while its actions have been enough to slightly destabilize the highly organized markets for decades, it is the market's margins that were large and some "downsizing" was only possible there through external third parties. A natural development of sorts, but not a revolution bringing along an epidemic.

Digital technologies have allowed a wider distribution of communications tools (mobile telephones, photographic equipment, sound recorders, geolocation devices, sensors for the state of health and well-being). The social progress linked to the power of wider communication is important, but a number cannot be attached to it. There are topical facts that illustrate this advancement through the distribution of reports shot on location, immediately resonating with official media and reaching a large audience. While society is better informed today, it also knows that it is more vulnerable as the spread of information is fast and significant.

4.6.3. *Company organization*

In order for a company targeting the transition to digital technology to succeed, it appears that its management should be organized in a collegial manner around few supervisors cooperating closely with the managers of the IT platform. The employment contract between the actors can be replaced by a tacit agreement between colleagues joining together for a quick and optimal profit without concerns over the time spent. The digital economy creates jobs that will not be allocated to

3 www.acpm.fr/Actualites/.

existing employees as a matter of course. Union demands must be forgotten as the digital economy is itself temporary, playing a game of cat and mouse with the administration and regulatory bodies. Wherever the digital economy may develop, the administration fails to enact legislation to stabilize the suggested approaches while respecting corporate and fiscal conventions.

A digital business is not managed in the same way as an ordinary business. It must be more watchful and faster in its decision-making, and because of this it must have indicators built in to its management software. On this subject, Stéphane Schultz, founder of "15 marches", a consultancy firm specializing in strategy and innovation[4], is addressing the management methods of young digital businesses and suggesting strategic plans to them [OST 10].

The commercial and financial management plan, or business model canvas, is a reference document that presents the way in which a business, or any public or private organization, plans to establish and ensure its profitability, the way in which it earns money, conquers parts of the market or increases its traffic. This fundamental document is based around nine components (source: 15 marches).

1) customer segments;

2) value proposition;

3) distribution channels;

4) customer relations;

5) revenue streams;

6) key activities;

7) key resources;

8) key partnerships;

9) cost structure.

4 http://15marches.fr/.

According to the authors, this functional decomposition should allow each element to be analyzed and all activities to be synthesized as well as possible.

The pace of innovation is currently so fast that heads of industry and trade are barely able to foresee or even survive the evolution of developments in digital applications. Their future is becoming hard to predict and it must be taken into account as the situation develops. Certainly, stabilizing the volume of use and a better control of applications could be achieved through increasing the cost of Internet usage.

4.6.4. *Digital refusers*

Who wants to profit from the digital economy? Alongside those fanatical about the innovations brought by the Internet, it must be noted that there are still fringe group of people who are either very reluctant to accept or unconvinced by the usefulness of these developments. The United Kingdom, which is currently making significant efforts toward the success of its "Digital Britain" program, bitterly stated in 2014 that an average of 12.6% of its citizens showed no interest in digital technology, with this proportion reaching almost 25% in Northern Ireland [UK 15].

Senior citizens, in particular, are slow to familiarize themselves with how to handle computers, tablets and high-speed applications. SMS exchanges are a step that many seniors have not managed to reach. There are fears (and not only in the United Kingdom) that digital connections between seniors and the administration will not be achieved. The absence of Internet cafes or centers for continuing education in home computing in rural areas, in particular, restricts Internet access among older people. Planners have forgotten the broad variety of user profiles.

5

The Impact of the Internet on the Economy

As information plays such a large role in our society, the Internet influences the entire economy as it affects a great number of activities. Network operators, the press, universities, the economy and politics must all take this new medium into account. Several common schools of thought on this topic are summarized here.

5.1. Network operators and the new economic order

5.1.1. *Governance and regulations*

5.1.1.1. *Arpanet, forebear of the Internet*

At the end of the 1960s, the Arpanet research project was launched to allow remote access to computerized documentary resources in the fields of science and technology. Following on from this, after other network experiments, the Internet was constructed around the idea that the network itself must be made only as intelligent as required to concentrate on point-to-point data transfer. Other concerns that were considered secondary at the time, such as security aspects, must be resolved through solutions linked to network access and terminals (computers, smartphones and tablets). It is actually easy to improve the periphery of the network, but it is very difficult to modify the heart of the network itself. Several proposals for international data networks were made before a definition of the Internet was reached. Since its public opening, the Internet has continually been developed as a test-bed network and provides users with new applications.

5.1.1.2. *The Internet, a scalable network testbed*

Several experimental networks have been built since Arpanet, with each having been designed for a specific purpose. In these subsequent experiments, it was found that the essential networks connected to scientific or technical databases were expensive, and that the unsatisfied demands of commercial traffic and the general public could be related. The Internet thereby found its vocation. It has also gradually opened up to high-speed multimedia services. The merger carried out through the IMS platform, connecting fixed-line networks to new mobile networks at increasingly high speeds, has ensured the success of the Internet in NGN, while simultaneously destroying old PSTN network structures.

At the moment, the Internet can seem inadequate (Appendix 12), but, gradually, initiatives to modify some of these aspects are appearing and new questions are being asked. In particular, the combination of the Internet alongside decreasing numbers of circuit-switched telephone networks in NGN raises questions of updating the principles of universal and public service (see section 4.2.2). Moreover, Internet usage challenges intellectual property, specifically copyright, unwanted content, unsolicited advertising and financial transactions outside State control. It is a long list and the fields of communication and related services happily overlap, sometimes calling the legal framework into question. On the Internet, multimedia, which, by definition, incorporates various audio, visual, textual and televisual means of expression can relate to several different regulatory bodies (ARCEP, CSA and SACEM for France). Texts relating to copyright, advertising and unwanted content must therefore be redesigned to be compatible with this new situation in the framework of the Information Society.

The Internet, taking its cue from governments and societies, follows a changing process rather than a static one. The Internet is used to respond to our needs and it improves as required. In the future, technical development will be necessary to support the growth in customers and connected devices, but it will also be necessary to change on an institutional level to respond to the needs and desires of States, which are still underrepresented in discussions on the development and governance of the Internet.

5.1.1.3. *The Internet, the network of cultural globalization*

Living without the Internet is today as unthinkable as living without electricity. In his novel "Ashes, ashes", René Barjavel imagines a situation where, devastated by the suddenness of society's break with technological abundance, citizens fall into social chaos, deprived of shared utilities and the most up-to-date information systems. In 30 years, the Internet has changed the way in which humanity communicates and interacts. Its gradual development has not followed any planned program in detail and the interruption of its operations caused a significant amount of public and private activity. In 2016, the advantages offered by the Internet are accompanied by new concerns about the balance of powers, equal access, centralization, respect for intellectual property and the security and global surveillance of the network in terms of materials and content.

Among the most urgent measures, it appears that Internet access must take into account the age and cultural sensibilities of Internet users. It does not seem acceptable that extracts from works as controversial as, for example, Adolf Hitler's "Mein Kampf" are put online. On the contrary, the development of the Internet should be steered toward the constructive advantages that this global network can offer humanity, such as for sustainable development, ecology, sustainable resources or saving bees.

5.1.1.4. *The neutrality of the Internet network*

First stated in 2003, the principle of "Internet neutrality" is that "anyone may access, without limitation, all of the available information and trade resources on the Internet". Despite (or because of) its naivety, this idea was behind a number of successful technical developments, related to several ethical, financial or regulatory problems, which continue to be unresolved. While Net Neutrality could support the traditional principles of public and universal service, conversely, it could also, nonetheless, collide with technological and economic constraints.

Mobile broadband, due to its own challenges linked to the necessary level of quality of service, requires a more complicated and aggressive network management than fixed-line networks. A 3 or 4G antenna divides the downstream speed between all users in the cell, which, for some of them, can at times damage some vulnerable applications. It is therefore necessary to restrict, upon entry to the network, some flows that could disrupt others by occurring simultaneously. On the other hand, Internet service providers and content providers must be able to balance their budgets and while some can profit from advertising, others do not have this opportunity.

Regulations must take into account the different technical and economic imperatives linked to the operation of the Internet. The European Commission is preparing a compromise text on the end of mobile roaming, which will come into force in summer 2017, and a new definition of "Internet Neutrality".

5.1.1.5. *Internet governance*

The transition of Internet governance from the American government to an international committee is the subject of a project available for public consultation, which was presented to the American Congress at the end of 2015. On July 31, 2015, the IANA Stewardship Transition Coordination Group (ICG) published its proposal that the role of supervising Internet addressing operations should be transferred to the global community. Since the 1990s, this has been taken care of by a body directly dependent on the United States government, the Internet Assigned Numbers Authority (IANA). The proposal made combines elements linked to the three categories of IANA operations – domain names, Internet number resources and protocol parameters. It anticipates that IANA's responsibilities will be entrusted to a new body dependent on ICANN, while a surveillance committee known as the Customer Standing Committee, which will have representation from various international players in the areas of industry, academia and regulation, will also be established. One of the objectives is to avoid a government being able to wield too much influence in this body. The new Governance should take into account the topics of health security (monitoring large epidemics, for example), guaranteeing security, respect for private data (not using big data) and combatting crime.

Traditionally, recognized network operators (RPOAs) bring together private actors and public administrations. During the 1990s, this shared interest group, introduced internationally in the 1950s, was supplemented by what have come to be called "high-level actors" (OTT), some of which are still not represented at the ITU. They provide original applications, both for a price and for free, and with or without advertising (Google, Facebook, YouTube, WhatsApp, etc.). OTT activity continues to avoid international and national regulation, and therefore also the taxation system to which national operators are regularly subjected. This asymmetry between new major communications service providers and traditional players in the life of the network will require international and European rules to be adjusted. Network operators would particularly like not to be obliged to provide OTT with fixed broadband connections at low prices, with the arguments thus far used in favor of investment in fact serving to conceal the creation of a competition, which has become unfair.

A single Internet, governed by numerous bodies and not only by the United States, must respond to everyone's desires. The governance of ICANN should remain transparent, open to all, and for this it should be entrusted to a large community, which brings together many actors from numerous countries. Faced with ICANN, the work of NETmundial is testament to the balance of power in Internet governance, in an environment scarred by the Snowden revelations.

Everything involving the client–server mode of operation on the Internet is the responsibility of the application provider, with the legal framework remaining within the area of Internet governance (therefore, in principle, to be split between ICANN and ISOC).

The majority of applications relating to P2P operation are covered by a legal loophole, subject to determining the responsibility of the author of the specific P2P software downloaded onto users' computers. In this way, transferring currencies in Bitcoin or securities in Blockchain should, in principle, fall within the criminal liability of the organizers of these applications. On the other hand, if this activity cannot be fiscally controlled by States, it cannot be prohibited. The situation is the same for the Deep web.

However, stricter regulations on the Internet could be gradually implemented following the agreements negotiated within international authorities (World Summit on the Information Society, or ICANN and IETF meetings). Even if the practical applications are slow to become effective, there are development channels worth considering in order to avoid the possible gradual fragmentation of the Internet, which would be damaging to everyone.

In principle, all inaccessible information on the Internet should produce an explanatory server response such as "HTTP 451 Unavailable for Legal Reasons". Following decisions taken jointly by the IESG and the IETF in December 2015, this response will henceforth be transmitted to the Internet user as "Error 451". In particular, when a site is blocked or censored by a government body, all requests made to this site will receive the response "Error 451", or in some cases "Error 404". The IETF anticipates that the government behind this censorship could accompany the error message with an explanation of the reasons for blocking access.

Sovereignty and security should dictate the Internet's fundamental rules. A report by the American NGO Freedom House indicated that, of 65 countries studied, China has one of the most restrictive Internet policies, ahead of Iran and Syria. It censors online content and blocks some Western sites, including Facebook, Twitter and Google. It believes that censorship is necessary to preserve the freedom of Internet users. "Freedom is the purpose of order and order is the guarantee of freedom. We must respect the rights of each country to govern their own cyberspace", the Chinese President emphasized, explaining that "no country should target online hegemony or interfere in the internal affairs of other countries".

5.1.1.6. *The French "Digital Republic" Act*

Following statements from the Council of State (*Conseil d'Etat*, www.republique-numerique.fr) and the French Competition Authority (ADLC), legal decisions will be taken in France in 2016. During the preliminary inquiry that proceeded voting on this act in January 2016, it was acknowledged that "the flow of data and knowledge" and "data of general interest (articles 10–12) could cause innovative services to be developed while contributing to the quality of existing services". The National Council for Digital Technology (*Conseil National du Numérique*) believed that public data service (article 9) would open up the possibility of "new social and economic opportunities". As for the second instance, on protecting rights in the digital society, the Council of State approved the insertion of the Neutrality principle (articles 19 and 20). A new legal framework for the Internet should therefore be implemented gradually in France and, indeed, in all countries, which would lead to some optimism for the future of the Internet.

5.1.2. *Advertising revenue*

In 2015, the economic model of Internet sites was still based on funding through advertising. This is at once a vulnerable, impenetrable and unbalanced model, as it is largely reserved for the benefit of OTT and providers of browser software.

Advertising is harmful to mobile Internet traffic as the displays are too large to be easily readable on smaller screens. Furthermore, to block invasive advertising, Internet users have the possibility of downloading specific software (for a price or for free), which deletes adverts from Internet page displays, unless they already have a browser that implements this function (Adblock Plus, being one of the most widespread). Reading information is easier and the privacy of the Internet user is a little better protected. For advertising professionals, such "ad blocking" could lead to a loss of around 40 billion dollars in 2016, which tacitly implies that, going forward, there will be a new organization in the financing model of all applications available on the Internet.

5.1.3. *Taxing the Internet*

The various Internet actors maximize their taxation systems in order to reduce the taxes that they could owe. The field's major players are in fact able to reduce the amount of tax to be paid to 35%. In this way, they benefit from the existing range of tax systems worldwide and from government apathy.

As an example, the amount of online advertising billed from a computerized platform in Ireland leads to a reduction in the consulting services provided to the parent company as well as the costs (salaries, social security charges, etc.). Ultimately, activity of close to 2 billion Euros does not lead to a tax of 17 million Euros. In Les Echos on November 24, 2015, Etienne Lefebvre pointed out that American multinationals use preferential international agreements to avoid tax burdens in Europe and repatriate their profit to the United States, thereby accumulating sums of over 2,100 billion dollars for strategic acquisitions from their competitors. In 2014, the GAFA companies saved some 170 billion Euros in this way, of which 600 million were at the expense of French finances. The international structures and agreements described in Chapters 2 and 3 serve large international companies more than they assist States.

Globalization enables this kind of activity and there are French network operators and actors in the Internet sector, which have moved their headquarters to Ireland or Guernsey. This behavior is not illegal as it does not contravene any international agreements, even if its facilities have only a marginal relevance to the economic and cultural expansion desirable for ensuring peace and prosperity in the world. It should be noted that the global regulatory "logic", which has led to the administrations operating national networks becoming private commercial law firms, has stripped States of revenue likely to help them fund their R&D activities relating to networks and communications services.

Faced with this situation, the OECD has begun the base erosion and profit shifting (BEPS) project (OECD/G20 Project) devoted to combatting BEPS. Solutions carried out through around fifteen different actions have been proposed by the G20 and the OECD to eliminate the loopholes that still exist in current international tax rules. The option of playing with "banking secrecy" and "tax havens" should be gradually reduced.

5.1.4. *Difficulties encountered by operators*

5.1.4.1. *Financial vulnerability*

Today, network operators are private companies and, as such, they can be the object of public takeover bids or be victims of bankruptcy. The versatility of the stock market and the possibility of obtaining significant loans offer high hopes to all those wishing to build large-scale economic empires. In June 2015, the Japanese billionaire Masayoshi Son tried to take control of the American operator Sprint in order to connect it to T-Mobile US, the third largest American mobile network, and thereby destabilize Verizon and AT&T, obliging them to respond to the financial market instead of investing for the benefit of their customers. However, this major operation unraveled due to the intervention of the FCC, the American regulatory authority. Nonetheless, this remains a risk for all actors and requires enough financial resources to be amassed to be able to face any potential danger.

5.1.4.2. *Operators and counterfeiting*

At the telecommunications network security workshop that took place in Geneva on September 15 and 16, 2014 at the ITU-T (ITU Workshop on "ICT Security Standardization for Developing Countries"), the delegate from Zambia explained the difficulties facing his country. Indeed, developing countries are punished by targeted attacks due to the shortcomings of their organization, the absence of a suitable legal framework for information technologies and the lack of technical expertise. The opening of Internet sites in Zambia has been immediately followed by attacks from hackers who have deleted the pages of websites, provoked by denial of service and spread through cloned SIM cards operating on mobile networks before being opened commercially. This was followed by significant financial losses for the three companies operating mobile networks and new web applications have been thrown into disrepute.

Investments from State and network operators have therefore not provided the development expected in economic activities, mainly because the needs of cyber security were underestimated on the legislative, regulatory and technical levels.

The Round Table at this Workshop showed that Zambia's case was not an isolated one and that the criminal industry, informed very early on of the value of the allotted radio frequencies in each country, is capable of quickly setting up large quantities of counterfeit terminals that do not have SIM cards and are therefore free to operate. To solve this problem, network operators are therefore obliged to pay for cloned terminals in order to be able to comfortably manage their networks [ITU 11].

5.1.4.3. *Unfair competition*

Alongside authorized network operators, in some States there are actors who offer the public illegal Internet access, which can also be network infiltration for espionage purposes. In Lebanon, a dozen countries have recently been identified as operating illegally in Beirut and its suburbs, through sensors placed in mountainous areas and pointed toward Turkey or Cyprus. The illegal Internet constitutes around 30% of Lebanon's Internet market. This unfair competition provides poor quality connections that are available throughout the country and are harmful to authorized operators.

5.1.5. *The slowdown of the electronics market*

The semiconductor market is a highly sensitive sector within today's globalized economy as, in recent times, it has defined the future of its own market. In practice, the microelectronic components industry is a private monopoly that fixes its own prices. Furthermore, it supports the existence of R&D in the scientific and military research sector [MUL 14].

The electronic components market seems to have been losing the dynamic that it had encouraged over the last 50 years and this change of pace is concerning analysts (see section 5.4.4).

5.1.6. *The four faces of hacking*

As total security is not the Internet's greatest strength, there are four feasible methods of fraudulent activity. The first category and its well-known variants, which allow everyone to spy on someone close or send them a virus they will barely be able to get rid of, can be skipped over.

The second mode of hacking consists of visiting unauthorized sites in order to be able to fraudulently watch and avoid paying for television shows (streaming). Close to 400 successful sites have been identified by the ALPA (the French Anti-Audiovisual Piracy Association), including those with limited audiences and legal platforms, such as YouTube, which can sometimes host illicit content. Some sites, such as "Zone Telechargement", with 3.4 million unique visitors, surpass the audience of French daily newspaper Les Echos.

Despite the risks involved, the third type of fraud consists of disrupting hackers identified by the grapevine. For example, Internet users aligning themselves with Anonymous have demanded the removal of almost 4,000 Twitter accounts "thought to belong to Jihadists" or connected to Daesh. The information may be surprising but it has not yet been refuted. Could Internet users be left to impose their own sanctions according to their own criteria without supervision?

The last option for carrying out fraud-related actions is provided by P2P Internet operation (see section 4.2). Besides sharing unauthorized music or video multimedia files, P2P opens up significant possibilities within Blockchain technology. Lastly, the Darknet, where Internet pages are not accessible through search engines and traditional browsers, provides the anonymity dreamed of by all potential wrongdoers (section 2.5.1.5).

The absence of an active administrator on the Internet network opens the way to multiple acts of fraud and many users reduce the number and duration of their connections for fear of potential hacking. The active and permanent presence of a "regional Internet watchdog" seems to be necessary.

5.2. The Internet's original provisions

5.2.1. *Communications packages*

With television output running out of steam, citizens seem to be turning toward a more useful and more constructive instrument that allows a greater cultural expansion and is available at all times of day. This connected tool allows users to escape, stay informed, improve their minds, educate themselves and amuse themselves a little at random at any time at the click of a mouse. The maximum amount of information from many sources is accessible in a minimum amount of time. In this meeting of Internet users and news sites, it is not strictly a matter of negotiations responding to the classic "supply and demand" model, but of a journey across the network responding to discerning or refined searches. This serendipity, or "finding something other than what was looked for", can also open the way to all kinds of risks.

For all network operators, established actors or new businesses, IP networks allow the expansion of all kinds of communications services in all countries, from voice systems to image services or 3D services, such as television or gaming. However, maintaining a network's financial balance remains, for the operator, a volatile and sensitive exercise.

5.2.1.1. *Specialized research networks*

Some scientific and technical information networks, whose existence is essential to university research and businesses, are described briefly below, as well others whose activity can be considered more or less legal.

5.2.1.1.1. *RENATER*

RENATER (*Réseau national de télécommunications pour la technologie, l'enseignement et la recherche* – National Telecommunication Network for Technology, Education and Research) is both a public information group and a French computer network. Since 1993, Renater has connected universities and several large research centers. With more than a thousand sites connected to its broadband links, the pan-European GÉANT2 network and the

Internet, Renater facilitates collaborative work between French researchers and their colleagues through multiple applications. Its infrastructure is continually updated to respond to user needs (www.renater.fr/).

5.2.1.1.2. GÉANT

The European Commission is developing connectivity in the countries in its Eastern Partnership, with a high-speed Internet network dedicated to research and education (R&E) through Armenia, Azerbaijan, Belarus, Georgia, Moldova and Ukraine. This will allow the interconnection of national research and education networks (NREN) to integrate them in the pan-European GÉANT network and allow researchers and students to fully participate in the digital economy. Two million scientists, academics and students belonging to over 700 institutions in this part of the world should benefit from this "helping hand" connectivity. Created over a decade ago, the GÉANT network (30 million users spread over more than 3,500 universities) has played a key role in assisting European research projects in fields as varied as climate change, environmental monitoring, energy supply, particle physics, radio astronomy, medicine and bioinformatics (http://europa.eu/).

5.2.1.1.3. Geonet project

According to the World Bank and the Oxford Internet Institute, in 2013 around thirty African countries had an Internet connection rate lower than 10%. The "Archipelago of Disconnection" covers three highly populous countries – Ethiopia, the Democratic Republic of Congo and Tanzania – where it is difficult for citizens to access information and culture. The Geonet project concerns the study of the knowledge economy in sub-Saharan Africa. More specifically, studies have been carried out to understand how communications technologies could transform the economies and the lives of working people in sub-Saharan Africa (http://geonet.oii.ox.ac.uk/).

5.2.1.2. *Hornet, the anonymous Internet*

The Hornet network, a joint project of the Federal Institute of Technology in Zurich and University College London, promises anonymous and secure navigation that is faster than the Internet. Currently, by downloading the Tor browser, it is possible to surf the Internet without fear of being monitored (see section 2.5.1.5 – Darknet and the Deepweb). But while the Tor connection covers a user's tracks, its browser makes the loading of webpages slower. Hornet, which improves Tor's operational principle, was used in 2015 by 2 million Internet users and is accessible securely through Facebook (http://arxiv.org/).

5.2.1.3. *The submarine network goldmine*

The global network of submarine cables plays a strategic role in Internet operation and therefore in the world economy. It is a particularly sought after boon for islands around the world due to the financial compensation linked to establishing GIX. Over half of the exits of the world's submarine cables are located on islands. Among the most fortunate are insular nations (Japan and Indonesia), the cities of Hong Kong and Abu Dhabi and the transportation hubs of Long Island, the origin of a large part of New York's land traffic. Hong Kong and Taiwan each have more cable entrances than all of continental China. By the same token, three times more cables arrive in Sicily than in Italy. In the Pacific, it is Guam that leads the way with 10 cable entrances.

In this regard, Nicole Starosielski, Assistant Professor of Media, Culture and Communication of New York University, underlines the digital potential of Polynesia on the global playing field. Today, the islands, which are politically and economically stable, are very attractive points for interconnection between continents. Some of these islands can thereby improve employment, and gain the subsequent economic resources and a privileged access to the global network. In Tahiti, the Honotua cable, laid in 2009, is underutilized, with 7 Gbps used out of the potential 640 Gbps, due to the high access costs, believed to be between 70 and 180 dollars per month, and because of the "digital cul-de-sac" effect caused by the absence of a second cable in the Asia-Pacific region positioned toward Latin America. A whole technological industry could emerge in this way, for example with a diversification of the investments geared toward China. However, the old balance of power established in the global computer network is, without a doubt, still not ready for such a change in strategy[1]. [STA 15].

5.2.1.4. *The Internet and cultural diversity*

At the start of data networks, around 1979, experts believed that the profitability of the infrastructure of networks, including Tymnet, Medline and Francis, would lead to them only being used by international research centers specializing in scientific and technical information, for the benefit of media houses, universities or financial management centers, such as Eurex. It was not until 1984 that the commercial opening of the Euronet network showed the interest of third party traffic ("third party" in the sense of "non-scientific or technical" traffic) on billable data networks.

In 2016, an Internet user with free time can access all available information (at times incorrect or out-of-date) and, based on their own knowledge and according to their tastes and concerns, form opinions for a modest price. The Internet is no longer just in the hands of engineers and nowadays everybody wins in terms of access, due to the power and availability of browser software and Wiki-style sites. The Web is full of pages that are as vulnerable as the works of great libraries would be in the face of fires, the devastation of war or un-referenced changes in content. In all cases, pages that have been bookmarked disappear over time, followed by reorganizations that can seem senseless. Nothing is set in stone. The most elementary rules for establishing contact, the dates of updates and referencing on the site and the identity of the site owner, are often forgotten. On the other hand, while the Web makes reference to the major classics of French and world literature, these irreplaceable pages are also juxtaposed with the outpourings of self-proclaimed Internet users, the authors of overly verbose texts and videos of varying quality. This digital cohabitation means that the participation of the smallest common denominator is inseparable from Web access and culture clash is therefore unavoidable. The laws of Kao and Odlyzko on the value of networks take on new meanings in this regard (see sections 1.7.2 and 1.7.3).

---

1 http://motherboard.vice.com/read/the-internets-undersea-infrastructure-is-due-for-a-disruption.

Social networks remind us, if proof were necessary, that people remain the same, despite these new means of getting in contact. They can free themselves from the constraints of distance to communicate even with strangers, and it is also possible to be in Paris and read the pages of Korean daily newspaper in French with the assistance of a free online translation service. Everyone can reap the cultural benefits according to their tastes, their available free time and their means.

5.2.2. *The press and information*

Well before 1980, the press called for a preferential tariff for their long distance communications from the Postal and Telecommunications Administrations. Regarding long distance communications, the press has long wanted the particular nature of their work, connected to the acquisition and dissemination of information, to be recognized. Their representatives expressed their desire for a significant discount, not only on postal tariffs, but also on the rental prices of leased lines, taking into account the need to have access to a dedicated network also designed for cultural and educational needs.

In the 1980s, a global network project especially designed for these needs was in fact planned, with its management to be shared between UNESCO and the Internet society [BAT 84]. Following numerous studies, this network project materialized in the form of the Internet in 1984. It is now available in most of the world's countries. The Internet provides, for a modest price, multilingual access to almost all sources of information, be it in alphanumeric mode, streaming media or on television, according to the subscription taken out or the data rate available. The paradox of this development is that this network, which was so ardently desired by the press, is leading silently to the destruction of its creators by reducing the number of their subscribers and advertising contributors.

The Internet, in fact, acts with the press, as it does with the creators of various artistic (music, films, video) or informational treasures. The pluralism of this output is therefore threatened by competition between the paper and digital versions, and, while users are delighted with the ease of accessing information, the number of jobs in newsrooms continues to decrease and media production is becoming concentrated in the hands of an ever smaller number of managers. In the United States, half of books are sold in a digital format and the proposed pricing level encourages this approach.

In 2015, the press still benefited from a special tariff for posting letters in bulk and sent within French territory. The press's conversion to digital technology could allow it to reduce costs and retain its independence. However, this transformation encounters two difficulties that are highly typical of this transition. Firstly, it requires expensive development as two operations, paper and Internet, must be managed simultaneously, while customers should gradually agree to migrate from one to the other. It must then develop a digital option acceptable for an occasional mobile Internet user and that can be used on different terminals (computer, tablet, smartphone or owner-operated e-reader), all while partially disguising some elements of the newspaper to attract new paying readers.

The different options that have been tried in North America and Europe (the Financial Times and the New York Times) show that the balance is quite difficult to achieve, even if the bold strategies are guaranteed to have the support of patient and understanding financiers. After 4 years of effort, the New York Times has managed to balance the budget of its double edition (paper and digital), bringing it a total of 4 million readers. Its management admitted that it had reduced operating costs, while the revenue brought in from newspaper sales had increased, although without succeeding in offsetting the decrease in advertising revenue. Newspapers are working on strengthening their digital presence, while the circulation of paper copies and advertising revenue continue to decrease.

Publishing paper copies of magazines is an activity with very high fixed costs that can reach, for example, 67% for a monthly title with a circulation of 100,000. The printing and transport costs are not directly related to the number of copies sold. For want of a secure payments system designed for small amounts and multiple recipients, the online press, weakened by strong competition from social networks, had not managed to exceed the 5% revenue of the written press. Uniquely, the regional and local press continues to exist due to their proximity to their readership. The written press therefore continues to receive State subsidies, even though its wish for easy access to a global network has been granted.

Online journalism is a new form of journalism that uses the Internet as its medium. The online publication of journalists' articles has sparked controversies relating to the copyright of these works. Online journalism takes the form of written texts, or texts presented in audio or video form on digital platforms. Digital journalism enables discussion and spontaneous commentary on current affairs at very varied levels, following the intervention of a more or less active and neutral moderator. It opens up the possibility of an "adjustable" kind of public expression.

While non-democratic regimes that do not respect freedom of expression present challenges for online journalism, a certain level of vetting should be exercised, although only to avoid anti-social behavior online. The status of an online editor is a long way from being defined internationally. Some experts believe that, because of the Internet, democracies risk losing their sources and their flows of information, as well as the plurality of their cultures [CAG 16].

5.2.3. *Adapting professions to digital technology*

5.2.3.1. *Digital technology involves entire businesses*

For some years, businesses have understood that the digital transition had a specific character and would take a long time to implement. The majority of them have already considered this topic or taken action. There is a multitude of ways to bring digital technology to a business. The IoT will be part of their revenge, as the battle will be played out in favor of information (data) and customer relations. The traditional players have a good command of the service equipment and the role of digital intermediaries is a transactional one. The whole issue consists of increasing the value of this information, to use it at the right time and in the right medium.

With online digital services, the commercial strategy is based on the principle of advertising paid for by good quality funds. One satisfied customer will bring in another. The curve of the level of satisfaction generally takes the shape of an inverse Gaussian distribution: few neutral customers and many satisfied and dissatisfied ones. Word-of-mouth on social networks is faster and more effective than large budgets dedicated to advertising spots on television. The reactions of the public are very fast and it is necessary to follow them with the appropriate technology. Connections between platforms open up vast opportunities, the consequences of which cannot be foreseen.

Digital technology is breaking up some professions as it leads to the establishment of new practices. As such, a kind of digital capitalism can be invoked, despite a continuing inability to determine its boundaries and aims. Music, the press, health, education, finance and real estate are areas that are already being transformed. Users are attracted by the availability and speed of access to services and, at the same time, they benefit from a large reduction in delivery costs and tariffs, even though the hacking of personal information remains a constant worry [ESC 15].

With smartphones and soon with connected objects and vehicles, physical actors are moving closer to value-added information and it is possible to build on this advantage in future economic models. It is essential to be able to respond to the following three main issues:

– managing the paces of different businesses as only one part of the model can be digitized;

– being able to integrate the best concealed innovations and skills as quickly as possible;

– integrating digital culture within the business, beginning with convincing the management, the CEO, its executive board and its Director of Computer Services.

Converting a business to digital is no small task. After several months spent digitizing individual processes, layers of interaction and validation must be added for a secure computing process. Only then does it become clear that the business requires a more fundamental change. This process is repeated time and again: one part of a task is digitized, but behind the scenes, the processes leading to it cannot be modified for reasons beyond the business's control. The advantages of digital provision in terms of cost reduction, operational efficiency and personalized customer service cannot be achieved. In practice, it is necessary to keep up with the latest technological trends, and the temptation to adopt them immediately must be resisted. It is important to innovate at a pace that suits the business, its market and its clients. It is equally important to take into account the modifications, which will need to be introduced, given the development of the field.

5.2.3.2. *The acknowledged pioneers of digital technology*

Digital technologies allow direct interaction with customers to take place in a faster and more transparent fashion, without only depending on intermediaries. Thanks to social networks, it is possible to know the impact of launches immediately, without having to wait for a report from a communications agency.

Digital action is based on strategic pillars. The first is the use of consumer data to measure the health of products and understand why the related talk is positive or negative. It is thereby possible to anticipate the peak consumption or suggest consumer sale points through geolocation. Subtler and more specialized forms of targeting are possible in digital technology than the traditional methods of market forecasting. The skills of individuals and of the management are also a major issue. All businesses are converting to digital and early adopters are rewarded.

5.2.3.3. *Innovations and productivity*

Digital technology improves certain aspects of productivity in industrial and commercial activity. However, in reality, the digitization of our economic and social lives has a highly overestimated impact on our capacity to improve production. Innovations are often closer to gadgets than significant technical advances. As described below, the progress of digital technology can only be compared with its own instruments or its own uses (see sections 1.8.1 and 1.8.2).

The digital revolution aims to replace workers with software. Digital technology allows knowledge to be better organized and networks enable knowledge to be spread more widely than in the past. The diversity of cultures and standards is at once enriching and a source of potential conflicts. According to the Solow paradox, the introduction of computers to the economy does not always translate into a massive increase in productivity. In reality, computer systems are complicated and require expenditure on maintenance and technical support [SOL 00].

The Gartner Group recently recalled that every dollar spent on "hardware", can require 2.3 dollars to be spent on advice, server maintenance or computer security. This proportion of additional expenditure has increased due to the technical complexity of the related equipment (accessories, robots, automatic guided vehicles, etc.). As well as these technical costs, there are geographically dispersed production costs and expenses linked to management technique, interactions and overall coordination.

> "On the one hand, management software packages lower the costs of accessing and sharing information. On the other hand, communications technologies reduce the decision-making autonomy of employees and managers by strengthening centralization. The introduction of ICT produces operational repercussions in the division of labor and the distribution of income. The freezing of productivity gains shows the limitations of interconnected technical systems. The negative externalities of the wave of ICT innovation explain, in part, the lack of synchronicity between the economic and social transformations and their interpretation in politics which can no longer guarantee the legal regulation of speculation and debt" [BER 12].

5.2.3.4. *Preventative detection of faults*

Car manufacturers currently use the Internet through ten fault alerts affecting cars already in circulation, which reduces intervention costs and improves the company's image. In fact, today's new models are complex and the "zero defect" standard is difficult to achieve in the factory. It is only when the vehicle is in service that the recurring defects appear.

5.2.3.5. *Digital technology improves lives*

The acceleration of international exchanges has made the innovation and expansion of digital techniques easier while guaranteeing relative world peace. The ITU recalls that these exchanges contributed to a reduction in the price of goods and services in the range of 30% between 2008 and 2011, allowing households in the third world to become better equipped. Following the invasion of data, the promises of digital technology are aimed at a better standard of living and more reliable health and hygiene using the tools and sensors of electronic health (e-health). The rods connected detect contaminated foodstuffs, while the scanner decrypts the molecular composition of an item of food or drink. The patient is reassured by the body energy sensor connected to the smartphone (quantified health). Medical information (or e-medicine) is available to all, with the advantages and risks that this can bring.

The performance monitoring carried out in businesses to manage staff and adapt them to their functions no longer has any meaning today. Continuous performance monitoring through IT allows precious time to be saved and mountains of paper to be avoided by quickly and completely transparently verifying the efficiency of employees through the tangible results obtained from their workstations. At worst, surveillance at the click of a button will be entrusted to Big Brother, but we can hope for the best. Who, in the hierarchy, would indeed dare to do away with contact with agents on the ground and thereby lose information essential to improving the company's organization? In order to succeed, digital technology must reduce the number of levels in the hierarchy.

5.2.3.6. *Self-driving vehicles*

Future driver-less vehicles are worrying the American police. This is because automatically driven vehicles, for better or worse uniting the IoT and artificial intelligence, can allow shadowing to be carried out at a distance and criminals could also use them to shake off law enforcement or "target" the police authorities. It must also determine where responsibility would reside in the event of an accident – with the driving assistance software or a fault in the vehicle's maintenance? On the other hand, farmers would be thrilled to be able to have driverless vehicles connected to satellite surveillance robots to ensure the well-being of their herds located in Brazil or Africa. However, determining the cost price of this supervision would require careful research.

5.2.3.7. *Difficulties for some companies*

The adaptation of businesses to digital technology cannot be taken for granted and serious mistakes can be made, involving significant investments, because the company's management did not foresee all aspects of this conversion, having instead imagined that the software would guarantee all necessary functions and resolve all difficulties connected to the new organization of services. Confusion also occurs when the management speaks of customer "needs" instead of a "desirable accomplishment". Changing the environment too quickly surprises both the customers and the staff responsible for sales with attitudes that the implementation plan may not have included in its assumptions.

Half of the businesses surveyed on this subject spoke of resistance to change regarding the transition to a digital style of management. However, the companies further on in the digital transition had a growth rate six times higher than companies lagging behind. The most stubborn companies can take inspiration from the aviation sector and the success of its BoostAeroSpace provision platform, which standardizes tools for collaboration between customers and providers. Bringing about the digital transformation of the economy is a requirement. It appears that this change is simpler when it relates to large companies that have had the opportunity to develop a project over a longer period of time than SMEs.

5.2.3.8. *Adapting the public sector*

Digitizing the public sector should improve the lives of citizens by simplifying procedures, avoiding duplication, improving the productivity of officials and saving the time of the two parties involved. If everyone is convinced by such a project, the transition of administration to digital action is not that simple, and from the very first step, all eventualities must be anticipated in order to avoid restarting the identification procedure during a citizen's second application. The "perfect" digital government must therefore predict all of the sectors, which would have need of citizens' cooperation in their administrative and fiscal management.

The British government has had several unsuccessful experiences with digitization (BBC programming, management of the National Health Service, pension services), which make it hesitate to launch a new project, despite having very quickly upgraded the national network for superfast broadband provision. The Digital Britain project, ironically dubbed "digital by default" by the British press[2], should be able to improve the lives of citizens and generate over 7 billion pounds worth of savings, but it largely encounters a lack of participation from British citizens and an overly slow digitization of public services.

France is ranked fourth globally in terms of digital administration. While it holds on to its place for use of its online services, the interministerial culture has difficulty following this development. Constructing a State platform is very ambitious. On this subject, Esther Duflo, a French economist, MIT professor and member of the Académie des technologies, recommends that a continuous trial suited to public policies take place in order to obtain assessments with concrete results. Digital technology provides extraordinary means to access the data that enables these assessments [BAN 12].

5.2.4. *The Internet and society*

5.2.4.1. *Social media*

Nothing has changed since Aesop, and "language is (still) the best and worst of things". Social networks are in this mold and their messages and videos amuse or irritate us in turn. The best rubs

2 https://www.theguardian.com/technology/2014/jun/23/when-the-uk-goes-digital-by-default-who-will-be-left-behind.

shoulders with the worst without it being possible to appeal to the reason or simple common sense of the thoughtless gossipmongers who freely throw around incivilities. The murderer proclaims his hate before setting himself alight in an attack and his message may be picked up during an incitement for compassion and world peace. The "global village" of social networks allows citizens to become closer to one another. However, as there is no filter upon entry, and culture shock can surprise at any moment.

At any given minute, social media is keeping all current subscribers to the Internet awake. However, a lack of respect for the minimum rules of decency has resulted in applications, celebrated upon their creation by an idealistic public, being thrown into disrepute. Following excessive enthusiasm, prudence and moderation are finally arriving.

5.2.4.2. *Digital technology and democracy*

At a conference organized on the idea that "the barbarians are attacking democracy", Nicolas Colin expressed his fear of an imminent systemic crisis [NIC 12]. Digital technology is attacking democracy, he says, because, by putting people in a network and giving them a means of direct expression, it gives the power to a chosen few who are difficult to govern, even though democracies are founded in principle on the concept of an "indivisible people". In France, digital technology, he says, allows particular interests to dominate public opinion. American democracy is, by its nature, based on a permanent balance of power between the interests of groups. While digital entrepreneurs are the first investors in electoral campaigns, the circle is complete, it seems. "The digital revolution facilitates communication in the public domain and aims to bypass institutions with pyramid structures through network organization". Some sarcastic tweets targeting jargon-heavy political messages enable the success of movements based on original opinions. The ability to react increases subscription to the applications that are closest to citizens, less expensive and better managed.

Cultural reactions still remain impervious to Internet standards and vice versa. It is indeed within this duality that problems with the

Internet could appear if it seeks to unify the cultural reactions of all Internet users from a distance.

5.2.4.3. *The Internet and pornography*

It is impossible to ignore one of the Internet's most profitable activities, pornography. On this subject, information is not skewed by hypocrisy or modesty. While the stated figures are stunning, they are almost all unverifiable. It is therefore generally agreed that a third of the Internet's daily traffic is made up of pornographic images, emanating from almost 12% of sites connected to the Web (being some 5 million sites, or more, probably). This traffic should constitute around 25% of search requests on the Internet. Of course, though the visitation records are not available in France, they are in distant countries where there is social censorship, the French public having been largely oversaturated since the creation of this technology through the erotic messaging service offered on Minitel (information which is also unverifiable). The revenue from this activity comes from remote sales of related products, particularly DVDs, as well as the hidden trade in address files. Everywhere that censorship is practiced (Asia, Arab countries), the State's turnover surpasses what which would be considered reasonable (between 50 and 100 billion dollars per year and per country).

Some approaches undertaken in support of global economic liberalism have had unexpected consequences that have, understandably, aroused the indignation of the writer Michel Houellebecq, who wrote that "liberal capitalism has extended its influence over consciousness; going hand in hand with it are commercialism, advertising, the absurd and snickering cult of economic efficiency, and the all-encompassing and unrestrained appetite for material wealth. Worse still, liberalism has spread from economic matters to sexual ones" [HOU 91].

In 2015, the Indian journalist Palash Krishna Mehrotra, shocked by the existence of almost 900 pornographic sites in India and wishing to defend women and children, wrote on the subject in DailyO, a New

Delhi newspaper: "Human beings are the same everywhere, but open societies tolerate differences while others sweep everything under the carpet. The Internet allows us to look at information and fantasy in turn. Sexual curiosity exerts a great influence on culture, but it remains unexpressed. If the Internet responds to this curiosity, it does so as a safety valve, or as a private means of sanctioned release. The opening of this blockage brings in billions of dollars in very different forms, never mentioned in economic cybersociological studies" [DAI 15].

The American non-profit organization "Enough is Enough", which is committed to making the Internet more acceptable for children and families, provides a wealth of relevant information on its website regarding pornography and pedophilia on the Internet, which is considered a pandemic (http://www.enough.org/).

Intended for readers wondering about the funding mechanism of free sites that bring in revenue, it should be noted that this financial paradox is based on a complex sector. This is guaranteed through a chain of intermediaries and brokers who compare people with different activities, from taking photos or making films to those that receive traffic because they have meaningful domain names and agree to monetize them by levying their contribution. The French computer expert at the source of this information points out that this chain of technicians is characterized by a comparative lack of ethics on the part of contributors. Indeed, a significant proportion of these websites spread "tacit malware" slightly "inadvertently", but it is possible to confirm that this function comes from a decision made by the site's manager. Furthermore, "as pornographic sites receive a lot of traffic, they are quite often hacked by unscrupulous computer technicians who install malicious codes and unmonitored rerouting systems. Various multiplication systems also enable the same 'click' to be sold to various brokers, in both directions. The IT associated with this activity enables the machines of customers to be analyzed, which increases interest in these connections as the results are communicated to other backers who are just as dubious as the first ones".

It should be noted that the highest resale price of a generic top-level domain (gTLD) name was ".sex", which brought in 13 million dollars in 2010.

5.3. The new economy

5.3.1. *Industrial and commercial cycles*

5.3.1.1. *Acceleration of the pace of innovation*

Digital technologies contribute to developing the end-to-end transfer of information. They support interconnection and interoperability and facilitate interdependence. They improve the coordination of economies, and the organization of businesses and markets. The Internet's calling is to be able to address a targeted public that is spread over a vast area. If it is a question of gathering together members of an association of boules players who are, moreover, poorly trained in typing on a keyboard, a simple notice in the village bakery will suffice. Conversely, if you are looking to bring together the parents of children afflicted by a rare illness, the Internet must be used. The technical progress made over the last 30 years has provided price reductions for all necessary commodities, including housing, healthcare and education. This craving for the best prices while retaining the level of quality of service, regardless of the location in question, still aims to accelerate progress. Social protection must become less expensive and the door is open to competition in activities based on digital media.

Throughout the history of capitalism, the traditional laws of the market have favored limiting innovation to ensure the income of the entire existing sector. Therefore, between 1950 and 1980, an industrial product lasted around 10 years, a period during which the company could carry out a production follow-up, maintenance, monitoring of patents in industrial property and an after-sales service. However, the economic paradigm has changed and the market changes faster today than previously due to globalization and the speed of exchanges between consumers.

In 2016, the tastes of the public are fixed around a commercial cycle lasting 3 years in principle. This irregular development should be taken into account by businesses as offers of services must be assessed for a shorter period and in a revisable way. The digital revolution is remodeling all sectors of the economy and redefining occupations. It is not a question of "creative destruction", one of the features of capitalism, but of an economic repositioning of the strengths of innovation in the face of means of communication, the main drivers of economic progress [SCH 74].

5.3.1.2. *Possible causes of this acceleration*

The acceleration in the pace of innovation was perhaps implicitly written into its DNA and it falls to experts and historians to explain why. In this respect, two examples from real networks may provide avenues for research.

Between 1920 and 1970, the driving principle behind telegraphic transmission were modulations (of position or amplitude) on analog telephone circuits, on the basis of a telephone circuit carrying 24 wire channels at a rate of 50 symbols per second. In general, every country had its own system and its own provider. On the border, the modulation in amplitude of the Belgian telegraph systems, for example, was converted into frequency modulation to enter France. There were no other original or uncomplicated technical solutions to propose and all parties stuck to their respective decisions, as validated by the ITU-T. However, this situation was a major disadvantage regarding the distortion accumulated depending on distance, with modems and switches each playing their part dangerously and threatening the correct functioning of the end teleprinter. The idea of an acceptable limited margin consequently led to the construction of the longest possible circuit carriers which were also, naturally, the least profitable as they cannot carry enough telegraph circuits.

In 1971 and 1972, digital telegraphy entered international standards, on the initiative of a small American company with relatively little instruction in the international field of transmission. Faced with this development, Japanese, French and British manufacturers presented their own proposals, although they had to admit the merits of the original plan and accept the legitimacy of the proposal. However, the young company ignored the meaning given to the polarities of connecting to the global network and had a case of bad luck, reversing the positive and negative voltages in its regulation. In the end, the principle of digital telegraphy was adopted and management of telegraph distortion is no longer a concern. Two similar standards coexist, of which only the second is used due to its compatibility with the international network. This standard has been written by traditional manufacturers (with the relevant patent rights). The innovation therefore entered into telegraph history sooner than anticipated because it was driven by a young, dynamic company that was not responsible for winding manufacturing plants and wanted to enter the world of networks.

A change of provision could benefit innovation. At the end of the 1960s, the first annual business directories appeared (Kompass), listing all businesses in France, both large and small, according to profession or domain. At this time, the first modems were available and an entire system of low speed (2.4 and 4.8 kbps) intercity information and telecommunications connections were set up to give structure to intercompany relations. The opening of the first large surfaces dates to this time and is explained by the expansion of the communications network and the progressive ease of exchanges between producers, packagers, transporters and merchants. The shift of the distribution sector's concentration, initiated by a grocer from the small French city of Landerneau, changed the habits of customers and distribution channels, and slowly brought about the disappearance of all small stores in the all villages, regardless of their expertise. The same movement continues today, although undoubtedly on an even larger scale, both nationally and internationally.

Progress begets progress and digital technologies are not solely responsible for the phenomenon of job destruction. As an example, the availability of light rechargeable batteries gave inventors the idea of developing small electric clippers for male hairdressing that are very easy to handle. This primarily simplified the lives of barbers, before contributing to the closure of barbershops. Unexpectedly, the tool changed hands and the sales of clippers increased. The ergonomics of the instrument led to its success.

5.3.2. *New economic paradigm*

Digital technology provides an opportunity to accelerate the shift created by monopolies in the economic world (see section 4.5). It is enough to have a computer platform connected to the Internet and equipped with robust client–server software and suitable memory to communicate with the entire world and offer low cost services, thanks to small free commercials and generous advertising sponsors. The new economy, associated with digital technologies, is rapidly transforming social and professional activity through the brutal competition it practices and the social disruption it creates. Everyone promotes their own agendas on the social economy that can be tinged with idealism. Well-known topics are recalled here and some of them are quite closely related to other subjects, such as peer-to-peer economies, sharing, innovation or “Uberization”.

According to philosopher Bernard Stiegler, transitioning to a “peer to peer” economy could become unavoidable and the market should finally fall in line with the logic of the sharing economy. This economic theory is based on the principles evoked by the strength of open intellectual property systems compared to closed systems. “Peer production is sustainable collectively, but not individually. Unpaid contributions are not enough for a person to live on. Peer production should join forces with capital in one way or another, at least until an alternative system appears”.

The Harvard University economist Dani Rodrik believes that free trade, however lauded it might be by economic theories and recommended in what is known as the “Washington consensus”, is not necessarily the fundamental factor required for growth as experience shows that alternative policies could prove more efficient. In particular, the aims of national sovereignty and democracy should not be overlooked [ROD 07].

Mobile terminals with Internet access call into question the assumptions of 20th Century capitalism. There is no need for offices, employees, an excessive hierarchy or unions said to be "representative". The qualifications of teams must adapt to the tasks proposed. A sharing economy, with freely agreed human resources, is being organized in requested environments, although its demarcation system is yet to be defined. The financial return is modest in itself, but it is sufficiently large to destabilize the traditional cumbersome systems already in place. Such sharing economies are only set up incidentally in service areas where the inherited margins are comfortable and where externalizing mobile computing can make the best use of all its resources, without traditional quality control procedures becoming obstacles. This economy is making particular progress in activities where managers have not taken the initiative to digitize their services. A way to simplify tasks was on their doorstep, but the hare left the initiative to the tortoise that benefited from the availability of funds loaned cheaply.

Clayton M. Christensen, a professor at Harvard Business School, has studied the consequences of innovations in commercial businesses and in society [BUS 97]. Peer-to-peer enables the creation of intellectual value, such as the encyclopedia Wikipedia or the development of Open Source software, and also of material goods such as vehicles (Sensorica, Wikispeed and Local Motors), provided that they promote a balanced system of exchange between givers and receivers.

It is no longer a question of an owned and centralized Télétel system, connected to the national Minitel system, but of extending this architecture to a borderless communications system. The digital economy is designed for a central site gathering technical and commercial intelligence for millions of loyal customers through the ergonomic simplicity of its communications interface, associated with a reduced regulatory framework.

5.3.3. *The typical actors in the digital economy*

5.3.3.1. *TheFamily*

TheFamily, a company established in 2013, is a European investor which assists and supports around one hundred young dynamic companies each year by offering them training and access to capital to develop activities relating to the "new economy". The founders of TheFamily, Nicolas Colin, a tax inspector, and Oussama Amma, develop the philosophy of this commercial approach on their blog. Instructors show new entrepreneurs how digital technology can quickly make traditional activities competitive. Even failures are valued by the team at TheFamily as a way of enriching group culture and being able to share a positive experience. In 2014, TheFamily, with the participation of the Caisse des Dépôts (the French Deposits Fund) and the Groupe la Poste (the French post office group), published a study entitled "The digital transition at the heart of business strategy" ("La transition numérique au cœur de la stratégie d'entreprise") [COL 15].

5.3.3.2. *The Barbarians*

These new economic players do not seem to shrink away from any process respectful of the values of the past and, for this reason are considered "Barbarians" by the literary and artistic world. In his work dedicated to this phenomenon, Alessandro Baricco analyzed the following strategy [BAR 14]. According to Baricco, the Barbarians use new technologies to popularize new uses for ergonomic software which captivates and appeals to customers, even when the overall quality of the services proposed is far from perfect. The contact is direct and without constraint. They make the novelty more accessible by meeting the standards of all social classes. "What does the network matter as long as the information or service requested is available". In the same vein, the American journalist Bryan Burrough referred to the tragedy associated with the takeover of Nabisco [BUR 07].

The Barbarian economy connects individuals, Internet users, contract workers available to carry out the services associated with an insignificant level of social protection and some freedom in organizing use of time. Blablacar, the international carpool company, and Helpling, a German platform for organizing domestic help online, are developing services based on these same principles.

5.3.3.3. *Rocket Internet*

Rocket Internet is a technology company, which is able to reproduce application models for taxi services such as Uber in other areas. It does not have magic algorithms, but rather specialists in incubators capable of cloning technological layouts for the benefit of young, dynamic companies. Indisputably, Rocket Internet is the most important Internet platform for creating digital businesses outside China and the United States. Even though the loss of its hundreds of subsidiary companies remains important, it is making enormous investments, particularly in e-commerce. In 2015, it employed over 30,000 people in 100 countries and was prepared to respond to the demands of over 5.4 million mobile customers. Well supported financially, the holding can survive with around ten of its largest and most dynamic companies. While increases in capital continue to come in, Rocket's engineers are perfecting the management and training software given to its affiliated companies. In two or 3 years, the group is sure to be in a position to repay its debts and distribute dividends to its shareholders. Currently, in 2016, in a "normal" world, the public would have cause to worry about the figures, but are we not now in the digital age?

5.3.3.4. *GAFA companies*

The oft-used acronym GAFA refers to the group of companies made up of Google, Amazon, Facebook and Apple, which are the Internet's most active and productive OTT bodies with an overall market capitalization between 200 and 760 billion dollars each as of April 2015. Large companies fear that these GAFA companies are developing direct links with consumers, bringing them down to the rank of mere providers with highly reduced margins.

With 1.4 billion users worldwide, including 29 million in France, the social network Facebook has information on its users and the events that they have spontaneously put online. The profiles of users are irreplaceable tools for Facebook's marketing service. The data collected can be used by Facebook directly or resold to other actors.

Facebook draws around three quarters of its revenue from mobile advertising.

Facebook wishes to push the popularity of its Messenger messaging service, which has surpassed 600 million users worldwide, even further. Facebook has also opened the use of Messenger to merchants through a tool allowing them to make their client relationships more fluid: order confirmations can be sent with the messaging service rather than by email. They can also communicate via this channel to tell their customers when their package has been sent, when it will arrive safe and sound, and so on. If there is a need, buyers can contact customer service through this intermediary.

5.3.3.5. *The role of OTT*

OTT (see Glossary) are actors, which participate in an alternative Internet multimedia provision service by bypassing network operators. The service provider is not able to control the content displayed or ensure respect for copyright. The service can deal with voice, SMS or video carrying via streaming through a social network (Facebook, for example).

Young companies with limited experience of operating a network are not the only ones to take advantage of the alliance of digital network resources combined with shared computing intelligence. Expertise-based alliances are also conceivable. In February 2016, a group of network operators and industrial partners was formed as the Partnering Operator Alliance, with the aim of sharing knowledge on "future digital services and trends" in order to succeed in marketing new applications in the wake of OTT.

A group of European countries, including the United Kingdom, has publicly stated its opposition to overly protectionist legislation on OTT, which France and Germany would like to put in place. The sales pitch put forward is based on the fact that "the single digital market can only become a success if it works harmoniously with innovation, investment and entrepreneurship. Regulations should only be implemented if there is a clear need". The European Commission should rule on this issue and developing countries are currently looking for solutions with the assistance of the ITU-T. It should be noted that, in February 2016, around twenty OTT were represented at ITU-T Study Group 3, which has seen some development.

5.3.4. *Factors for success provided by NGN*

The new digital economy is developing on the Internet due to the emergence of several convergent factors. On the one hand, communications tools are now readily available and, despite the variety of their operating systems and the diversity of the mobile display systems of terminals, NGN ensures, in both fixed-line and mobile services, the compatibility of transactions through platforms connected to the Internet with global coverage. The significant increase in the number of multifunctional mobile devices (smartphones) is an important factor in the success of NGN, as it is for applications offered by the Internet.

The fabric of connections established through the different layers of the Internet (the surface visible to the general public and the Deep web for insiders, as well as for hackers) enables an extraordinary sharing of knowledge and exchange of ideas, with the best at times rubbing shoulders with the worst. Innovation emerges from these chance encounters between young discoverers and daring investors, fearless and indifferent to disapproval. Low interest rates and cheap loans are encouraging the launch of products and services, which would have been unthinkable in the past. The Internet is also a sector for exploring sentimental and commercial urges, in spite of cultural differences and the continuously disputed regulations put in place. As the pace of creativity increases, the political classes are concerned by social changes, increasing unemployment and the lack of certainties in this development. Like Pandora's Box, the spread of new information technologies has become part of all societies and has prompted an entire range of reactions. The tragedies caused by Daesh can probably also be considered signs of rejection in the face of the technological progress following on from the international trade liberalization measures organized by the United Nations. Traditionally, the filter of history will certainly only remember the best of this progress.

There should be neither dreaming nor complaining. The sharing economy has known limitations as it has its own limitations as a medium. The Barbarians have only attacked parts of limited markets, and which raise financial and social questions. In the United States, in 2014, self-employed people with no employees only made up 8% of total employment (a proportion which remains stable). Big business has understood the lesson and is pondering its response.

5.4. Longevity of network services

5.4.1. *Political and regulatory variables*

Networks worldwide are currently positioning themselves toward building NGN which bring together the Internet, IP protocols and fixed-line and mobile networks. De facto governance of the Internet presently resides with ICANN and the international group of financial and regulatory bodies described in section 3.1. As all States accept this situation, the principles held by this technology serve as a framework for the services and applications of 70,000 Internet networks, as of April 2015. Even though some isolated Internet networks can be set up, it seems difficult to imagine a State which refuses to connect to the global Internet.

However, amendments to NGN are possible on the regulatory and technical levels and some countries have attempted their own, limited, experiments on this issue. Regarding security and the confidentiality of personal information, the European Union could certainly have acted in a more vigorous and tangible way than it did in the wake of the PRISM scandal. Within technical standardization circles, several delegations are expressing their desires for a development more in line with the needs of their populations. Indeed, NGN which, as indicated, at once comprise traditional telephone networks, fixed-line and mobile networks and the Internet, are rich in potential services, although the diversity of uses makes it difficult to have a single "governance" applicable to a wide variety of audiences. On the other hand, the continuous development of the Internet is surprising decision makers through the unpredictable direction of the uses it can support. This explains the delay observed in regulatory decisions which at once affect both network operators, through penalties imposed retrospectively, and, indirectly, users. Because of technological developments and the emergence of new practices, the regulation of digital communications must evolve, although it can only do this late and by adding new considerations to an already complicated situation.

5.4.2. *Technological risks*

5.4.2.1. *Shortage of rare metals*

Computer equipment, just like devices for mobile telephony, use components made up of around sixty chemical elements and rare metals. These include critical non-renewable substances, such as antimony and tantalum, which will become difficult to procure by 2020 and 2090, respectively. So-called "blood minerals" (tin, tantalum, tungsten, gold) are extracted under very difficult conditions and are sometimes a source of funding for armed groups in African conflict zones. Mining resources are limited by the energy cost of operating them, and by pollution and its social impact [BAR 15].

In 2010, the Dodd-Frank Act made it necessary for American companies to reveal whether their production lines involved the use of "blood minerals". From 2016, this act will be strengthened in order to oblige businesses to provide more specific reports about the origin of their supply of provision. The European Parliament recently adopted an equivalent measure. As a result, an increase in the cost of terminals and computer equipment is to be expected. Modifications to current technological sectors and greater durability of terminals are also part of potential scenarios.

5.4.2.2. *Miniaturization of components*

The components industry is increasingly approaching the physical limits of microelectronics, referred to by the term "the Wall", at which point transistors would consist of nothing more than some atoms and insulation. It is already necessary to develop new methods in order to reduce consumption of computer equipment (Climate Savers Computing Initiative, a study evaluating performance in FLOPS per Watt). For instance, there are plans to store three-dimensional transistors, and to use nanotubes in molecular transistors, DNA computers, photolithography in the X-ray frequency band, and quantum computing, that is computing using "nano-computing" technologies. Tomorrow's microelectronics could move away from silicon in favor of new materials, with new materials for chips.

The profitability of new generations of machines depends on a future which is, at the very least, uncertain and it is conceivable that, under these conditions, it could be an economic decision and not a physical blockage that puts an end to the rate of development seen over the last 50 years in the components industry.

The limit of profitability of manufacturing new generations of components seems to be linked to the acceptable noise level and energy consumption, which leads to a freeze in the level of performance, in order to ensure the market profitability of more varied microelectronics and to reduce the volume of waste produced by the incessant renewal of generations of equipment.

5.4.3. *The Moore Conjecture*

In 1965, Gordon Moore, Director of Fairchild Semiconductor, stated that the density of electronic chips doubles every year and this trend could continue in the future, using the economic promises of electronic integration and bringing about, stage by stage, the development of all consumer electronics [MOO 65]. In reality, manufacturing the first Fairchild integrated circuits was much more expensive that the traditional methods of assembling separate elements mounted on a circuit board. Mass production justified dynamic manufacturing technology and, conversely, this allowed production to be carried out at a low unit price (Figure 5.1).

Gordon Moore pushed the investigation further and determined the economic optimum for manufacturing components in accordance with their number over a substratum, with this parameter enabling assessment of the profitability per unit of manufacturing components for the next generation achieved through photolithography. In reality, for reasons of competition between companies, Fairchild positioned the developments of a generation of components on three concurrent axes: correcting the faults of the previous generation, doubling the capability within the allotted space and improving performance (all for a sales price barely higher than for the previous generation). Indeed, it appears that the cost of an industrial product decreases between 25 and 30% every time that the cumulative quantity produced doubles.

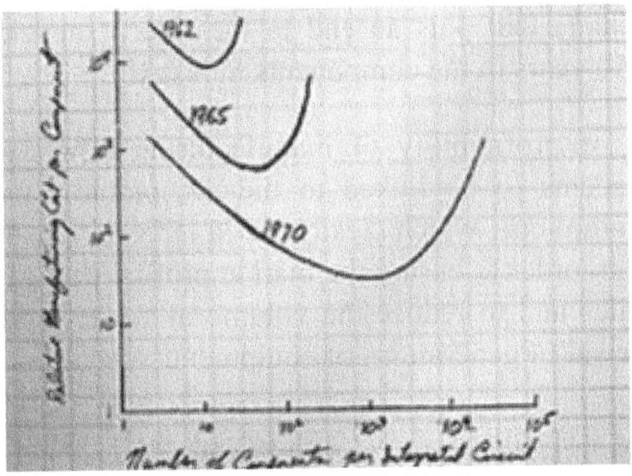

Figure 5.1. *Economics of producing integrated circuits, according to G. Moore (On the x-axis, the number of circuits per integrated circuit. On the y-axis, the relative costs of manufacturing components)*

This arrangement thereby allowed more components to be added in the same space, as well as reducing the volume of even more efficient components, correcting the omissions and mistakes in the previous versions and, lastly, making the numerous pieces of equipment used by manufacturers and laboratories much less expensive [IEE 15].

This "Moore Conjecture" (as opposed to "Moore's Law") has provided a very useful framework for prospective development studies for other sectors of activity (petrochemistry, medicine, telecommunications, engineering, etc.) for 50 years. It serves as a basis for financial regulations for investing in components as it brings an economic advantage to an industry supported by a solvent market with high demand. Figure 5.2 shows that the revenues from the global semiconductor industry have indeed regularly developed at a ratio close to 100 between 1968 and 2004.

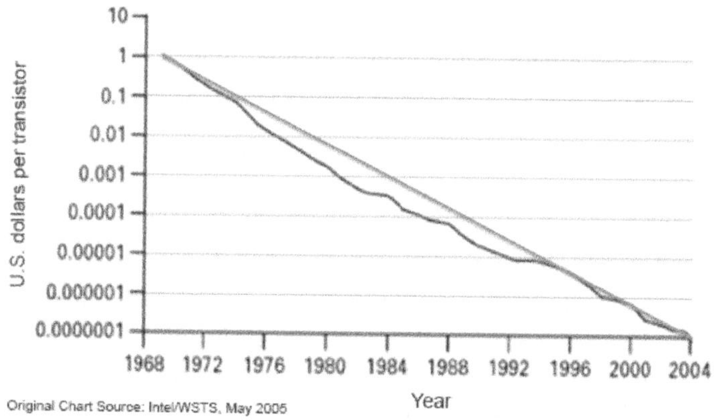

Figure 5.2. *Revenue of the semiconductor industry between 1968 and 2004*

In 2015, the Moore conjecture seems to have switched to a rate of two and a half years unless it is completed, due to lithography costs having become prohibitive, the heat released, the physical limitations of components and overly large leakage currents. However, it would be daring to connect the economy to technology and claim that global GDP would be able to maintain a mathematical relationship with the Moore conjecture, given the numerous factors likely to arise.

5.4.4. *The crisis of the components industry*

5.4.4.1. *Decreasing demand*

Since their inception, semiconductor factories have continuously operated 24 hours a day in order to respond to growing public demand and to attempt to make production profitable at the lowest possible cost over many years. Since 2008, consumer demand for silicon has started to decrease due to the reduction of salaries, which are no longer in line with productivity. This downturn has led to a reduction in investments from semiconductor heavyweights in silicon wafers 450 mm in diameter, which provide the most potential revenue for the industry. At the end of 2015, it seemed that the growth of the global semiconductor industry could not make progress in producing 450 mm silicon wafers due to low consumer demand. Lacking customers, monopolistic capitalism dictated that development had slowed by itself.

The growth of the economy is dependent on its capability to produce materials in the largest possible quantities and to take advantage of more and more salaried consumers. Yet the pace of overall productivity in Europe is only slowing down. Today, the field of consumer electronics is dominated by American and Asian companies. The relocation of this industry from the United States to Asia has opened up a significant rift in the American commercial balance and led to a subsequent depletion of employees in the electronics sector.

R&D in semiconductors and pharmaceutical products is investing increasing amounts without producing innovative results. Twice the amount of capital is required to produce the same amount of goods as 50 years ago and, while these services make life easier, their role in economic productivity is almost non-existent. Demand is still insufficient for stimulating fruitful investment. Big Data and Cloud Computing technologies provide information that is admittedly interesting, but whose consequences are limited with regard to productivity [ART 15].

The increasing cost of production lines is driving manufacturers of microelectronics to consolidate their investments. The inflation in the costs of product development also favors these transactions, which enable businesses to grow more quickly. In 2015, almost 70 mergers and acquisitions in the components sector reached a hereto unheard of level of frenzy with the total value of buyouts exceeding 127 billion dollars, and the phenomenon of outbidding the competition could increase further in 2016. With costs rising faster than revenue, manufacturers are called upon to invest further in designing chips, developing manufacturing processes and production equipment. The transition to sharper burning is increasing the costs of developing chips and the manufacturing processes are growing 30 to 35% with each new generation of technology, while the revenue of the semiconductor industry has been increasing by less than 5% per year on average since 2010 and threatens to decrease by 2% in 2016.

5.4.4.2. *The 10 nano sector*

Intel, like the Korean Samsung and the Taiwanese TSMC, is obliged to postpone the release of the generation of 10 nanometer chips until the end of 2017. This delay will push back the arrival of the next generation of 7 nm burnings, previously planned for 2018, but which will now probably not appear before 2020. The 10 nm semiconductor technology is associated with the introduction of extreme ultraviolet radiation into laser lithography.

5.4.4.3. *The end of the Japanese miracle*

Professor Yotaro Hatamura of Tokyo University maintains that his country's industry fell victim to the sin of pride in giving itself the title of "permanent technological powerhouse" in the 1970s. However, while Japan's economic development over this period has been obvious, the quality of the technology in Japanese products has been associated with very high prices. Japan is being rapidly surpassed by low cost and medium quality Chinese and South Korean manufacturing. For instance, with regard to components, in Japan, the manufacturing of semiconductors only tolerated a coefficient of 5% in defective items, whereas for the Korean Samsung, the rejection rate could reach 30–40%. However, in cornering the global market through very low prices, the Koreans have been able to gradually improve quality and subsequently move up to first place in innovation.

5.4.4.4. *The Chinese strategy*

China occupied 56% of the global semiconductor market in 2014 and recently toughened its policy on buying foreign digital equipment. Some American providers are excluded outright. China's ambition is in fact to strengthen its semiconductor industry in order to reduce its dependence on imports. "Made in China 2025" aims to move toward

40% self-sufficiency in integrated circuits by 2020, then to 70% in 2025, which can only contribute to destabilizing the global market.

5.4.4.5. *The Indian market*

The current equilibrium is precarious as India, tired of living with a significant unemployment rate, decided to move into manufacturing electronic products in its "Made in India" program in order to improve its trade balance and reduce the poverty rate. The Indian approach will undoubtedly provoke a reaction from China and probably a change of political strategy in its relations with the United States. At the same time, industrial plants which currently participate in the inexpensive manufacturing of counterfeit or poor quality products will turn toward honest production as it is paid on a more solid footing [BAT 13].

One possible path for Asia would be to invest in government-sponsored semiconductor manufacturing plants, with agreed investments receiving guarantees in sustainable capital financed by external and internal funds with a view to an open international market for developing countries.

5.4.4.6. *The market of the IoT*

The expected El Dorado of the market of the IoT also compels providers to complete their product portfolio as quickly as possible. The components of the IoT are a specific technology related to producing fully depleted silicon-on-insulator, with lower power and at a much lower cost price. The IoT should amount to a potential market of 7,000 billion dollars in 2020, on the assumption of 50 billion connected objects, although the market of standalone microcontrollers for the IoT barely surpassed 400 million dollars in 2015 which still does not really constitute a growth catalyst able to cope with the downturn in the components sector. Furthermore, the anticipated jackpot should be shared between several competing transmission systems (see section 2.7.12).

On the other hand, the energy requirements associated with the IoT pose an environmental problem. Moreover, there is a risk of conflicting radio messages in the IoT, which could cause network saturation. The dangers of radioelectric emissions to human health have still not been conclusively proven (Appendix 4).

The energy consumed can be produced through carbon-free sources, but a means of storage without environmental damage must be found and thermal dissipation should also be researched. Low-energy Bluetooth enables battery use to be reduced. Objects with positive energy are worth considering with reclamation of all energy microsources (movement, light, heat), subject to profitability.

According to ABI Research, the global market of the IoT should reach 5.4 billion connections in 2020, higher than the 1.2 billion devices connected today. Every business, every house and every school should therefore have its own local network connected to a larger network, all matched to routing, protection and addressing systems that the global components industry will find difficult to finance without significant funding. It is no longer the Moore connection that clearly shows the outrageousness of the IoT project, but the banking network which is questioning its financing. Would the rival World Bank established in China be able to refuse loans without the caution proposed by the United States?

5.4.4.7. *The prospects of the components industry*

Over the last 50 years, the integrated circuits industry, which is related to electronic activity, has been the information age's development engine. In 2016, it appears that the era of the personal computer, the driver of the semiconductor industry, has come to an end. The smartphone has picked up the baton, being responsible for 27% of electronic component sales, even though this market is in decline. Vertical integration allows the margin payed to chip providers to be maintained internally and the design of its terminals to be better controlled. For IC Insights, these operations show that buyers are suffering from a lack of growth in their traditional markets, which is pushing them to expand the scope of their activity through acquisitions in order to continue satisfying investors. According to ABI Research, vertical integration could affect half of the smartphones sold worldwide from 2020. R&D costs are increasing by 35% at every stage and financing for university research is decreasing.

Asian and European R&D is increasing its efforts while the United States is falling slightly behind.

Demand among American consumers has not been greatly stimulated by the financial measures taken under the guise of "quantitative easing" in 2008 and, for want of anything else, financial institutions have lowered interest rates, which has contributed to speculation and exacerbated unemployment in developed countries in Europe and in the United States. The change in the interest rate should reassert the value of the dollar to the detriment of other currencies. The monetary policy of the United States should therefore be reviewed in order to revitalize internal demand and consequently increase employment. The global components industry prepared a roadmap for 2015–2030 within the ITRS framework (International Technology Roadmap for Semiconductors) and it is to be hoped that the decisions taken fit into the framework of intelligent macroeconomic policies. The transition to silicon wafers 450 mm in diameter is essential to improving the profitability of mass production and to be able to confront the global economic downturn. Without immediate reforms in this area, a resurgence seems impossible and the significant slowdown in components activity will push the American economy toward an inevitable depression (http://apekmulay.com/)

5.5. The Internet and politics

5.5.1. *Monetary policy in 2016*

The state of current affairs confirms that, in 2016, the economy had weak growth and low inflation. The consequence of this is seen in the always highly accommodating nature of global monetary policy. This situation is increasingly leading to significant economic and financial irrationalities [SOW 06]:

– boosting financial assets often disconnected from their fundamental elements;

– monetary creation that does not correspond to credit;

– negative interest rates in the short sections of some rate curves;

– unjustified very long-term investment on the economic and financial levels;

– overvaluation of financial assets over a globalized geographical area;

– asymmetry of responsibilities in a system mischaracterized as liberal.

In order to govern according to its tastes or the current real or presumed needs, political authorities always use the resources provided by the economy. The acceleration of the pace of growth has endangered the future of our society, and a new economic strategy which complies with the minimum moral values seems indispensable. Past economic theories only partially responded to their era's most noticeable problems. We have the advantage of being able to make the most objective possible critical reading, without giving in to passion or despair.

It is to be regretted, as the philosopher Jean-Pierre Dupuy said, that economists do not have an overview of the wider world in its entirety [DUP 12]. "It is an understandable human concern to seek to push oneself ever further, but it seems to support dangerous myths. Should the artificial intelligence of our most up-to-date computers be called upon, paradoxically, to assist political authorities in resolving the problems posed by management of the global economy?"

5.5.2. *The Internet, an instrument of capitalism*

All telecommunications activities are, to some extent, placed under the control of the institutions described in section 3.1 whose operation seems directed more in favor of developed countries and the companies dominating the Internet market (see sections 5.3.3.4 and 5.3.3.5). In order to respond to China's repeated demands, the IMF introduced the Chinese currency into the SDR (a unit of account), a symbolic status, but one which will have obligations for the transparency of the Chinese financial markets from the end of 2016 (section 4.3.1.2).

Some experts believe that the IoT should promote the emergence of a sharing economy, no longer based on capital but on the exchange of goods and services, which would change the direction of capitalism. Furthermore, with the IoT, the effectiveness of people and of systems should be increased and a "society with zero marginal cost" should appear, thereby resolving all of the planet's resource problems, including climate change. Such is the enticingly naive Épinal image which, it is to be hoped, contains some element of truth.

For others, contrarily, democracy is threatened by the Internet. Industries will continue to be dominated by it for a long time and the IoT is close to a technological myth. Capitalism is built on the concept of scarcity and, very fortunately, progress has been able to push its boundaries. The Internet is not killing capitalism as it possesses data, and this is maintaining its power.

A new topic of concern is taking shape with the announcement of the TTIP (Transatlantic Trade and Investment Partnership between the European Union and the United States) and CEFTA (free trade agreement between Canada and the United States) international treaty projects. These two documents (still being drafted) aim to fundamentally change the content and processes of developing public standards, by broadly integrating the interests of multinationals to the detriment of general interests and conflicting with democratic choices. Particularly affected by these treaties are new public services yet to be defined, the definition of common standards on services, the data protection environment, the protection of intellectual property, the dispute resolution procedure between the state and businesses and the creation of a new private supranational jurisdiction, which bypasses national and federal public jurisdictions through private arbitration tribunals, or ISDS (Investor-State-Dispute Settlement). Paradoxically, this issue brings together highly concerning legal texts and relatively low economic gains (less than 0.4–0.5% of GDP). The mystery surrounding these draft regulatory developments should, in principle, be resolved through diplomatic meetings which should take place after several presidential elections:

> "The stakes concern the survival of democracy if the market is global, but democracy is not. The rules of the

market prevail over the rule of law, and regulatory bodies do not take unemployment or deflation into account. However, education, security and health fall within the duties of states and democracies are not able to impose their rules", suggests Jacques Attali, on a different note.

5.5.3. *The Internet, an instrument for domination*

The Internet's first decade supported the idea of a participatory system that would encourage the primacy of representatives to the detriment of all pyramid-shaped forms of authority. A "junior Internet", made up of young students, foreshadows the authorities being called into question by self-organized, supplied and independent groups. However, public intelligence has not been able to take charge of the network, undoubtedly because of the awkwardness of the technology, and the movement and abundance of actors. While the Internet enables the development of some collective intelligence, limitations eventually appear as soon as it becomes a question of the practices of the greatest possible number of people. Wikipedia does not meet all the criteria set out at its establishment and many collaborative projects remain exceptions. In order to last, collaborative forms are based on relatively strict arranged processes. The Internet should therefore proceed according to the indicator that its Governance Forum will give it every year.

Cyberspace at once concerns both the public and private domains, all while being subject to the actions of economic actors, including those associated with the Barbarians, the GAFA companies and OTT with multiple resources, strongly supported by the United States government. These large bodies contribute to the economic standing of the United States and they use the weakness of Internet regulations to their advantage, while European countries have still not defined a common digital policy.

Commercial domination is primarily practiced through the software required by large actors. One of the strategic strengths of Google is its Android mobile operating system (in Open Source), which was used by the majority of mobile devices and tablets in service in 2015, and which is found in television sets (Android TV), computers (Android-x86) and connected objects (Android Wear). Google's omnipresence is mimicked by Apple, with its iPhone terminals, the ApplePay mobile payment system and in its Watch devices.

At the end of 2015, the Internet connected over 3 billion users around the world. Its technical expansion also focused on developing its governance. Social groups or cross border communities participate in trade, placing the international network with no designated owner in competition with the national networks of states. The regulation of the Internet is at once based on both technical aspects (who takes decisions on expansion, who maintains the network, who receives and redistributes money?) and on content (pedophilia, pornography, personal security, counter-terrorism, intellectual property rights, respect for personal data). The content and uses of the Internet are largely unregulated and, consequently, there is a willing tendency to move toward state regulation, which remains controversial.

India intends to build its digital applications itself and it does not wish to freely squander the personal data of its own Internet users. The Indian regulatory authority, TRAI, refused to allow Facebook to set up a package of applications (Free Basics) on a social network tailored to the Indian market. The dishonest paternalism of Internet heavyweights is intolerable for developing countries and especially for Asian countries that, through their own culture, have a different concept of the principle of Internet Neutrality.

The Internet Governance Forum is the group that is responsible for establishing common goals each year, in order to implement them according to the regulations of every state and the international agreements that have been signed [FRE 12]:

> "Who currently dominates the Internet?" Quite clearly, those who take part in ICANN meetings and the Internet Governance Forum may reply. While the United States has accepted some compliance with these bodies, they have eloquently expressed the idea that they would remain the only state to sit in on these decision-making

groups in order to enable decisions to be taken quickly. The United States sits alongside universities, research centers and all bodies able to provide elements to make technical progress in the network and to enable the services most useful to all of humanity to flourish. The ascent of emerging countries (Brazil, China, India, Japan and Russia), faced with the economic downturn of the United States, is subject to comments on the inflexibility of these states. Indeed, China represents 13% of global GDP and 30% of the GDP of the developing world by itself. The concept of "regulation", necessary for proper management of the Internet, still needs to be defined by all participating states and, subject to a faithful translation, be accessible to all human cultures.

5.5.4. *The dangers of the American economy*

The United States represents 23% of global GDP and 12% of trade in goods, yet around 60% of global production lies in the dollar area. The part of the stock of international investments belonging to American companies has decreased, currently reaching 24%. Today, the managers of American funds on Wall Street manage 55% of the world's assets. The growing gap between the economic and financial power of the United States is creating difficulties for other countries and the costs of the dollar's domination are beginning to win out over its advantages. Other countries put up with the risks of the New York Stock Exchange and have no protection in this area. The United States is extending its financial influence ever further to complement its political actions. The Euro and the Chinese currency (the Yuan or Renminbi) are still not in a position to rival the dollar, but the American monetary system seems to have reached its limit and its collapse would have global consequences.

5.5.5. *The worries of the West*

The slowdown of the Chinese economy is restricting international trade and putting the economies of emerging countries at a disadvantage. On the other hand, the rise of an aggressive form of radical Islam is surprising and terrifying the West. “After the fall of the Soviet bloc, the Western world appears incapable of establishing a safe international environment which is favorable to the longevity of its values and way of life. The emergence of Islamic terrorism is accompanied by the collapse of structures in the countries these terrorists have come from. The Islamic world, made up of 1.6 billion people, or 23% of the world’s population, formerly the most peaceful possible area, has become one of the most violent and this violence is expressed, through the arrival of refugees, in the heart of Europe and even threatens the prospects for the recovery of the global economy. Commercial expansion is currently falling behind and the political order favorable to globalization is gradually wearing away” (free translation of a text by Robert Skidelsky, professor of political economy at the University of Warwick).

5.6. Experts faced with the future of the Internet

The IMF regularly readjusts its growth predictions for different parts of the world and pronounces a set of recommendations. It indicates that, for 2016 and 2017, “global growth could decrease if important transitions in the world economy are not well managed” [EMI 16].

As the conclusion of this essay on the economics of networks approaches, it would have been desirable to provide a vision of the future, reassuring if possible and supported by the logic of the facts presented in the preceding pages. However, overall predictions seem quite uncertain. The opinions of the experts consulted on this topic are scattered between a generous and poetic idealism and slightly surprising visions of an apocalypse. The truth probably sits in between these viewpoints, by maintaining our trust in a gradual adjustment of networks in response to the needs of their users.

5.6.1. *The sharing economy*

5.6.1.1. *Resourcefulness*

Summing up, Monique Dagnaud, a sociologist and founder of the application OuiShare, said that "the so-called 'collaborative' economy seems an economy of resourcefulness". The collaborative economy is focused on the sharing of goods and services (renting rooms, carpooling, computer training, zero-rate loans, organizing local events, intellectual output (studies, publications), accelerating projects, training, sharing Internet access), by ignoring social protection and political conservatism. Its trade is balanced in theory in peer-to-peer social relations, due to the decrease in buying power and according to the resources developed by different generational profiles [DAG 13, FIL XX].

5.6.1.2. *The associative framework*

However, the different experiments that have been attempted could only be a passing phase in the development of our society. Declared associations aligning themselves with the French 1901 Act (Loi de 1901) or "social cooperatives" are at once in competition with and the saviors of this sharing economy, particularly with regard to families in need and people with disabilities. Family associations were already very active between 1945 and 1980, despite highly reduced means in the framework of mutual assistance, in organizing useful pastimes for young people and renting knitting machines, but they have been transformed by the Internet, expanding their audience and adjusting to the new needs of households. Today, family structures are splintered and the motivations for action are organized around more targeted and individual goals, which, despite the distance between partners, the Internet is able to respond to even when its users are scattered.

5.6.1.3. *The "Uberization" of the economy*

"Uberization" refers to the adoption of the trading model described in previous paragraphs, consisting of, in the short term, making resources available to customers through their smartphones, the availability of the Internet and geolocation resources in every pocket. The Uberization of work was already the subject of research in 1995 (including the Boissonat report and the work of legal expert Alain Supiot on social drawing rights). It also inspired the draft law on creating a personal activity account, which should enter into force in France in 2017. Under the cover of responding to the still hazy goals of the so-called "collaborative" economy, client–server platforms are raking in profits that are energizing the brokers at the Stock Exchange, attributing a stock market value of 40 billion dollars to the global taxi firm Uber in 2015, nearly as much as network operator Orange. However, in reality, as Henry Curty observed, "with Uber, Airbnb, Ouishare, Ouicar, the Hive that says yes, it is not a question of a collaborative economy. These companies have a transitory nature and do not have assets. These are elements of neoliberal capitalism at its most raw" (http://thierrycurty.fr/).

New players arrive through customers, where the barrier to entry is weak and in a line of business where innovation is reduced. Digital businesses are inventing a new customer experience, which is rapidly snowballing through the network effect and the management of Big Data. Among the victims are libraries, travel agencies and dealerships. Digital technology is thereby causing companies to tip into hyper-concentration: the big get bigger and the weak fall by the wayside. The world of decreasing returns has given way to one of increasing returns. Customers bring in ever more other customers. In this context, economists and politicians believe that it would be appropriate to define a regulation fitting "digital workers", with suitable social and fiscal provisions. As the Uber phenomenon is sustained by unemployment and the reduction in the revenue of average households, these factors encourage the kind of part-time work associated with the flexibility provided by the Internet.

5.6.2. *The social justification of the Internet*

Apek Mulay, an economic expert, notes that it took 45 years for the telephone network to make its way into the majority of American homes and that, while the Internet has been three times as fast, there are still 4.4 billion people worldwide with no connection. The price to be paid for transporting voice signals has been very high since networks were created, and today, the expected prospects of the IoT raise many questions. Will the progress promised by these new technologies be geared toward the profit of the large industrial and banking groups that rule the world or will it rather be designed to further increase the charges imposed on citizens? Innovations, which often come from small dynamic companies, will increase the profits of large international groups. The question naturally arises of how to know whether there is reason to extend the debt financing of a global economic system, which is not designed for the benefit of those who take part in it and which, on the other hand, exacerbates the unemployment rates of all countries. The globalization of the economy should transform it into a collaborative form of capitalism where the ecosystem would distribute profits to all instead of gifting its fruits to a select few [MUL 15].

5.6.3. *An economy of inequalities*

For Thomas Piketty, a social science researcher, wealth accumulation is growing even faster than earnings, and income inequalities are largely based on land, real estate and professional capital. The price of land and accommodation is increasing specifically because of progress in ICT, with limited quantities of physical goods more valuable than "digital goods", products that are available in large quantities at decreasing prices: ideas, knowledge, software pictures, games, videos, and so on. The financial system feeds into speculative debt by channeling investors toward high-yield assets. ICT development allows markets and financial centers to interconnect and investors have the opportunity to borrow funds quickly. As a result, inequalities are growing [PIK 13].

5.6.4. *The ebb and flow of capital*

The global economy depends largely on massive movements of capital and, in particular, those of the reserve formed of excess global savings, a large nest egg set aside by some economies. "The reversals of trends", says Hélène Rey, an economist in international macroeconomics and professor at London Business School, "are often linked to decisions made by the American Federal Reserve (Fed), due to the dollar's dominance of the global economy. The risk return and the price of assets changes with the markets as soon as the Fed makes so much as the slightest decision, to the extent that a "currency war" is sometimes referred to, with the weakest economies paying the price for the consequences of these abrupt fluctuations". Indeed, in opening up their capital markets to finance their economic growth, some countries are losing control of their economies [REY 15].

5.6.5. *A financial catastrophe in the making*

Ravi Batra, an American of Indian descent, is an associate professor of economics. In several of his books on the topic of financial capitalism, he has explained the global financial and economic crises, which have occurred since 1750. On the basis of this historical analysis of capitalism, Ravi Batra shows that humanity goes through cycles lasting an average of 35 years, with inequality generating economic performances and social changes. Each cycle thereby sees successive movements of anarchy with the predators of the financial circuits, the reestablishment of authority through the military, democracy with intellectuals and universities, and then the arrival of power in the hands of corrupt individuals, which gives way to a new predatory anarchy. This cyclical view of history in four periods is enlivened by comments explaining the occasional doubling of the length of the cycles (the communist regime that lasted 70 years) and the shift of global economic centers from London to New York.

Over the last 40 years, Ravi Batra has made several economic predictions, 95% of which have become reality within a few years (the fall of communism, the Iran-Iraq war, the 1980s stock market crash, and so on) [BAT 78]. He believes that capitalism will self-destruct due to its following a path with no purpose other than to obtain ever more money, regardless of the crises and collapses, while the number of poor people only continues to increase [BAT 07]. He has proclaimed a "rapid and superb" recovery of the American ideas based on democratic principles that should spread around the world, allowing poverty, unemployment and income inequality to be eradicated and finally bringing about the "golden age" that has so often been heralded. His most spectacular declarations include:

> "Employment finds its source in monopolistic capitalism, which allows there to be an industrial and commercial gigantism at exorbitant costs, creating enormous profits and limited rewards".
>
> "Truth always prevails" and "honesty, integrity and ethics are the real jewels of every individual, rather than money".
>
> "Large macroeconomic changes are to come in 2016 and the semiconductor industry must be ready for such changes. Businesses that could not adjust to changes in the manufacture of semiconductors are doomed to disappear" (http://apekmulay.com/).

Since 1981, the trade deficits of the United States have gradually weakened its industry, particularly the semiconductor and high-tech sectors. Ravi Batra pointed out that, while the United States practices a policy of free trade, Japan and China do not play along and prefer to intervene in the exchange market in order to reduce the cost of their currencies in relation to the dollar [BAT 15].

Since 2008, the present economic crisis has moved toward reducing trade. Workers see their resources deteriorating while deficits increase. The question arises as to how to reduce the social burdens of companies in an era where computing and communications systems allow there to be reductions in the time taken to adjust or to distribute new products and services. The response given to this question is to abandon the traditional regulatory framework and place the business into a new framework, bringing less restrictive regulations into play and thereby reducing the costs of producing and distributing services using computer platforms for mass distribution, to the detriment of workers' rights. In the United States, this "proletarian army 2.0" represents 5 million workers, programmers, content editors, editors, graphic designers and video effects specialists. These repetitive and alienating tasks have not been included in the labor code [GAL 83, ACE 13].

5.6.6. *The economic cycles of electronics*

Nikolai Kondratiev (1892–1938) was a Russian economist who launched the theory of economic cycles that supports the idea of the world economy undergoing alternative periods of growth (improvement and prosperity), followed by periods of recession and depression. These macroeconomic cycles could be explained by the inertia inherent in the awareness of the need to invest in order to restart consumer demand after a period of depression [KON 26]. On the other hand, creating money restarts demand, but an excess of it can cause financial crises later [DEN 66]. Recently taken up again by Indian and American economists, this theory has been expanded with examples of economic cycles linked to the steam engine and the cotton industry (1800), steel and railways (1850), the electrical and chemical industries (1900), the petrochemical and automotive industries (1950) and, more recently, information technologies (2000).

Detailed analysis of the global semiconductor market in 2015 revealed the existence of difficult situation, which risks becoming exacerbated further. The discrepancy between excessive production of semiconductors worldwide and the overly low wage incomes in the United States is likely to cause a recession, especially since China and India are getting ready to take on some independence in this area (see section 5.4.4.7).

5.6.7. *Moving toward an economic "third age"*

Between 1994 and 2007, several countries experienced severe financial crises that worsened the situations of millions of people. Decision makers and economists have questioned the pace and sequence of deregulation and liberalization measures. International financial bodies have therefore altered their practices. The "Washington consensus" (2008), regarding the adjustment of measures taken in favor of globalization, has not been universally popular among economists or Nobel Prize winners in economics. It has been roundly condemned by anti-globalists. The need for a strong state and for state intervention to reduce poverty have been recognized by the World Bank. The IMF has tacitly authorized states to use their budget to increase their social budgets, including by opening up a deficit in order to avoid a depression. The IMF recognized that free movement of capital endangers the economy, requiring a certain level of regulation and, if required, of taxation.

For the Canadian economic and scientific essay writer Jeremy Rifkin, the economy is based on three pillars: transport (or logistics), communications and energy. With the Internet and digital technology, producing goods and services no longer costs anything, apart from the initial cost. Even then, this zero marginal cost concerns cultural goods (music, books, etc.) rather than renewable energies and physical products, which are assisted by 3D printing and distance learning. This zero marginal cost economy could sound the death knell of capitalism. The Internet of communications, the Internet of energy (with smart grids) and the Internet of logistics are coming together within the intelligent infrastructure of the IoT. That this system is free harms the profits of businesses and property rights, as an economy of shortages replaces the economy of plenty, which has sustained us. In short, the IoT could lead us to a planning movement in the framework of a third industrial revolution, as hoped for in northern France, with the advent of an ecological world in the COP21 mold, by reducing the length of the working week to 30 hours and with a social contract connecting the highly skilled jobs of less than 20% of the population and other low skilled and low paid jobs. These proposals could find a place in a Third Age of humanity by developing the IoT, cooperative businesses and empathy through social communicators [RIF 14].

This new industrial revolution is the product of the eventful association of various technological advances, including artificial intelligence, robotics, the IoT, self-driving vehicles, 3D printing and biotechnologies. Large businesses will organize themselves in a more decentralized and horizontal manner by creating platforms, flexible networks and small independent bodies. This revolution is destroying jobs in intermediate transaction functions. Considerable social tensions are at risk of appearing, transforming the economy and overwhelming the traditional concepts of democracy.

5.6.8. *The end of the world and the death of the Internet*

On several levels (political, agricultural, industrial, energetics, social, environmental), our society must confront significant problems, which have no logical solutions. History tells us that, in similar cases, regardless of the initial cause, people rise up against institutions and blindly destroy all that they believe to be responsible for their distress, before starting over. However, for the first time in history, the crisis is not restricted to a single country but is global. For want of the power to imagine the details of the impending future revolution, three strands of thought are taking shape to predict that which we know to be impossible – the end of the Internet.

5.6.8.1. *Moving toward the collapse of the financial world*

Banks and markets are collapsing. The global economy is threatening to fall into recession. Banking supervisory authorities are multiplying. They have dispensed with ready cash without taking the potential risks to the economy into account. Accounting standards are continually being revised. The role of rating agencies has not been clarified. The consequences of the massive creation of currency have not been analyzed. A currency war is brewing within the great global economic imbalances. A new Bretton Woods-style international agreement is becoming necessary, with unambiguously worded goals for a new period of at least 70 years.

Some experts have taken the decision, and the risk, of alerting us as the global collapse of the economy is now accelerating and the end of the world is, it seems, near. The mythical figure of Singapore trader Gordon Gekko is evoked to remind us that global stocks are overvalued, that debt is reaching its apex and that financial reserves have disappeared. In short, as the charts provided in support of this thesis seem to confirm, we have been on the brink of a global financial catastrophe since the beginning of 2016 (see http://kingworldnews.com/).

5.6.8.2. *The end of the Internet in 2023*

Another pessimistic point of view has been put forward by Andrew Ellis, a professor at Aston University in Birmingham, which is based on observations of the Internet's real traffic. The continuous development of data exchanges is not without risks, as it could produce an irreversible flooding, causing a breakdown in the functioning of the Internet ("Broadband Capacity Crunch"). Sooner or later, the demand for broadband will catch up with the feasible supply of global network routers, even in PON-OLT fiber-optics. Professor Andrew Ellis has announced the date of the Internet's ultimate collapse as 2023. He concludes by hoping that the political classes decide to increase Internet access costs to reduce demand, or to reduce the access time and data rate per user.

5.6.8.3. *Power failure*

"Wrong!" cries another expert, a specialist in microelectronics. "The Internet will break down by 2020 anyway for want of sufficient electrical power". With consumption growing at a pace of 40% per year, Internet demand for electrical power could indeed soon catch up with global production capacity. The electronics components expert Bernd Hoefflinger describes a series of scenarios that would allow the breakdown of the Internet by 2020 to be avoided and guarantee that the sustained growth of nanoelectronics continued beyond 2030. Indeed, the world needs to take up many radical changes in the future and some of them deserve to be taken into consideration now [HOE 16].

For each of these aspects, the easiest conclusion to present to the reader is to reference a cataclysmic scenario that no one could possibly believe. Admittedly, many factors emphasize the difficulty of changing the world of electronics and computing. Human society has always managed to cope with the challenges of the new proposals and technical adjustments made possible by science. We have never been able to predict how the world will change but, whatever the future brings, we will once again be able to adapt to any problems.

6

The Economy and the Future of the Internet

6.1. Building a national network

The initial establishment of a national communications network is never guided by key economic principles. Just as past actions permeate the present, so do they also influence the future. Networks are built according to the different pressures (political, financial, normative, customer oriented) that they face and their construction should be forward looking. The various theories built around network growth do not permit the formulation of formal laws, in correlation with compelling economic evidence. Furthermore, countries with liberal economies, such as socialist economies, have, after experiments of varying lengths, abandoned the principle of planned interventionism, a Colbertist or Marxist tendency. There are also few rules able to assist network managers in carrying out their duties. Building a network consequently responds to the simplest criteria possible in the liberal world: taking the best decision guided by the available technical achievements, with the selection being made by network operators and validated by the good will of business and residential users. If no basic rule can be provided to assist in constructing or developing a network, economists will be frustrated. They will then turn, by default, toward establishing indexes relating to the use of digital services, which will allow them to observe the degree of user satisfaction in all countries over the course of a day.

6.2. Internet network structures

In 2016, as previously, a national communications network consists of a call collection network and a transport network. Traditional fixed networks and new mobile networks, which are experiencing very strong growth, are connected on the basis of IP protocols, through digital technology, and both have access to the global Internet. This group forms the NGN. More often than not, users forget about this recent architecture, built by connecting several networks, and they remember only that they can access the Internet through numerous applications available on computers, smartphones and mobile tablets.

There are three possible modes of operation on the Internet: the widely used client–server mode, which operates through questions and answers exchanged using a specialized server; the P2P mode, which makes a peer-to-peer connection in isolation; and the Darkweb for those users who like to visit its hidden world of mystery and danger. The number of applications and intermediary actors on these digital network structures is increasing. The rapid development of technology surprises even experienced actors and it makes short-term provision difficult. New network architecture provides more than predicted, pleasing NGN designers and shifting the center of interest of most applications toward mobile devices.

Overall, the majority of traffic is generated by outsiders to the world of network operation (OTT), which currently acquire their revenue through advertising and by reselling client profiles. Not only is the major profitability of the network eluding established operators, but the content and volume of the traffic actually being exchanged is not being monitored by anyone. The supervision of structures and the quality of service have been abandoned to the extent that further questions are beginning to be asked about the technical governance of the Internet. Universal interconnection undoubtedly requires a review of the principle of non-hierarchization of the networks that make up the Internet.

6.3. Network regulations and pricing systems

In much of the world, network operation is included under the services regulation system, which complies with the agreements concluded between various international bodies and states, in the spirit of the "Washington Consensus". Respect for shared technical standards effectively facilitates communication.

The regulatory and pricing framework of communications services has been adapted to NGN structures and favors players specializing in innovative applications, OTT, and companies that bypass established and new network operators. The regulatory authorities take action to avoid positions of dominance developing between operators in the wholesale market, but it is highly sensitive for them to intervene in the retail market.

The number of subscribers and the volumes of traffic exchanged have increased considerably in a relatively short space of time. The operating margin is decreasing while network expansions are becoming necessary. Existing regulations, which focus too much on a policy of low tariffs, establish strong competition between operators, favor OTT, weaken network operators and reduce their flexibility.

6.4. The issue of supply and demand

Following a long period of shortages in methods of communication, businesses and the general public currently have a wide choice of ways to communicate, born of technological advances and the efforts of skilled commercial businesses, including OTT. The latter aim to respond to the real or implied needs of potential customers and they sometimes suggest playful innovative services to their customers as experiments. Businesses specializing in distance selling and banks continue to play close attention to these developments and the related security levels.

One of the most important features of the global structure of the Internet network relates to its capacity to receive a very small number or a very wide range of simultaneous calls per day on its server platforms. While this network flexibility has its own limitations, it offers information providers immense opportunities at very low prices. In particular, all Internet actors have the opportunity to respond to demand from very different parts of the market, such as the "long tail" segments shown in Figure 4.1, which minimize management costs.

Remote computing has been developing for around 50 years, and today the numerous digital activities that it has brought about surprise users with their originality and their rapid and silent growth, in both fixed-line and mobile networks and for all audiences and in most fields.

While innovation at all costs stimulates the development of many new applications, it also encourages the potential risks related to transaction security with Blockchain, still poorly understood, and the approved (and free) traffic on the Darknet. This abundance of new, seemingly uncoordinated, proposals casts serious doubts on the success of projects related to the Internet of Things, and on the quality of service and good level of security in digital communications.

6.5. The Internet and the economy

The existence of a global network as widespread as the Internet, with its three billion users, naturally has a significant cultural and economic impact with no historical precedent. Information is transmitted from one part of the planet to another in a very short time, accelerating all kinds of responses in the modern world. There can no longer be a question of the "network economy" when 70,000 networks are interconnected under the same protocol and are able to transmit multimedia messages to disparate terminals. The cost of a network is nothing compared to the economic and cultural benefits provided by these interconnections. For a marginal cost, essential information can produce phenomenal results. Who could evaluate the unit cost of a kilobyte for a service provided? Even experts refuse to pass judgment. Eloi Laurent, a researcher at Stanford, declared on this subject that "there are no truths in economics. There are only assumptions ahead and choices behind" [LAU 16].

Yet the press, which was one of the historic originators of this global network 60 years ago, has still not managed to master the digital technologies at its heart and reduce its own distribution costs. The transition to digital is indeed also difficult for businesses and, even more so it seems, for administrations.

However, it is clear that the roles of actors who participate in this network and services architecture are quite badly assigned, as the operators take on message conveyance to the great benefit of OTT, who largely avoid taxation in the countries where they operate. The economic downturn favors the entry into e-commerce of intermediaries (the Barbarians) that overturn traditions and call national professional regulations into question. At the same time, the very fast growth of ICT and the electronic components sector have sparked fears of a risk of serious technological disruption. Will we be able to accommodate all challenges related to the development of the Internet?

The global impact of the Internet is so important that the question arises as to whether it could become an instrument imposed in international trade to serve the interests of a specific country. While some question the role played by the logic of digital technology in relation to the foundations of capitalism and free trade, others are interested in the potential consequences of short-term developments. If the growth of the Internet is related to developments in globalization, then is this network not also an instrument intended to change the basis of capitalism and reduce cultural diversity?

6.6. An Internet of optimists

In any case, with NGN being the Internet's carrier, the whole world has at least basic knowledge of the global, multiservice and multilingual network, available in fixed and mobile situations. Everyone can understand its advantages and disadvantages. Solutions to resolve the various technical issues mentioned previously are being researched and there is a growing clamor to amend the "Washington consensus". OTT, which are beginning to participate in ITU-T standardization work, must learn to cooperate with network operators. Private companies are already proposing to co-finance some network structures in the most disadvantaged countries. In order to be able to extend its growth with new offers, the Internet needs more appropriate regulations and financing.

On the Internet, it is difficult to reconcile innovation, freedom of expression and respect for personal data while at the same time protecting Internet users against malicious attacks, deviant eavesdropping, incivilities and aggressive commercial dynamics. Tim Berners-Lee recently made a plea for a universal Internet charter, which would aim to protect both human rights and innovation. The paradox of the Internet is that, while it is possible to encrypt messages, they can then most often only be read by specialists. If the content remains inaccessible, then it is the State that opposes this form of communication. Freedom of expression is caught up with this mystifying box. If the Internet's official structures are unable to provide a solution that satisfies all parties, the fallout will gradually be felt in the least developed countries that, rightly, believe that they have more to lose than to gain in this unbalanced exchange. Nothing will be able to prevent these countries from simultaneously having two Internets, one closed and the other open.

Since the First World War, the potential collapse of our society has been repeatedly evoked. The 1972 Meadows report pointed to the physical limitations of the humanity's resources and the 2014 Intergovernmental Panel on Climate Change (IPCC) report focuses on the climate problems facing us in the 21st Century. Unbridled growth defines itself by its own limitations. Dialog between states should allow new rules to be determined worldwide and priorities to be set. Ecology should guide the economy, which, using the means available, will determine the social norms to be respected through the fairest possible distributions, while avoiding the use of "blood minerals", respecting the environment and appropriately managing rare resources. Would it therefore be possible, without doubts and in a more stable world, to draft reasonable and justified "economic laws" on the topic of communications networks which are consistent with individual freedoms?

To conclude this essay on a note both more concrete and more scientific, let us observe that the recent expansion in telecommunications has been behind large discoveries made in the exploration of our planet. Alongside space exploration, the positioning of communications satellites has given us the opportunity to study the makeup of the upper levels of the Earth's atmosphere. Similarly, the requirements associated with positioning submarine cables have enabled the biological community to learn about the wonders of the deep. The constraints associated with coordinating its former colonial empire obliged France to set up various submarine connections and obtain, over the past 40 years, some deep-sea knowledge. France has also taken on the responsibility of managing the world's second largest exclusive economic zone (11 million km^2) that spans three oceans, a task that can only be carried out using the international legal and technical framework for coordination. The potential shortages of some critical metals justify the diversification of supply sources and the research of their potential. Utilizing these shared deep-sea assets requires the abilities to monitor aquatic surfaces, make use of the biological resources in their depths and determine how to manage aspects such as food potential, the presence of fish or radioactivity and respect for the environment [DES 10, IEE 15].

In some ways, the Internet conjures up an image of its potential that continues to obscure the depths and responsibilities related to its operation. All states are at once users of the Internet and responsible for managing it. The principal global economic powers have sought to take stock of the available land and sea resources and make use of them, and, since 1945, the United States has been one of the main drivers behind this policy, developed by the United Nations with the assistance of relevant financial bodies, within the framework of a large humanitarian mission. China, for its part, has constructed a mass interventionist capitalism based around a reduced set of standards and actors, including operators and manufacturers. A reduced economy of focused and effective means gives rise to low-cost services. The union of these two types of dynamics should allow the pressing questions currently being posed to be answered, namely:

– improving the network of networks so that it ceases to be a source of profit for a small number of preferred companies;

– making full use of deep-sea resources by no longer viewing the oceans as a shared dumping ground;

– combining our efforts in all areas in order to save the planet.

Appendices

Appendix 1

GNP, GNI and GDP

Gross national product (GNP): is an economic indicator, devised in the 1930s, which corresponds to the wealth produced by a country's inhabitants and expatriates over the course of a year. It allows the wealth produced by a country to be measured. To make this assessment, the GNP takes into account the value of the goods and services created and subtracts the value of goods and services lost or altered during the production stage. In contrast to gross domestic product (GDP), which measures the wealth produced by all operators and persons residing in a specific territory, the GNP is calculated on the basis of a country's citizens, regardless of their place of residence. However, this indicator has not been used in French accounting since 1993 and has been replaced by the gross national income (GNI).

Gross national income (GNI): corresponds to the total income (salaries and financial revenue), as understood by national economic operators. The GNI is the sum of the GDP and the balance of the major revenue streams from the rest of the world.

Gross domestic product (GDP): is an economic indicator that aims to quantify the total value of wealth (goods and services) created by economic operators within a given territory (households, businesses, public administrations). The annual variation in the GDP reflects the rate of economic growth. The GDP is a monetary flow that can be calculated by combining the value added by all players involved in production (including administration), either by counting up all salaries, profits and financial revenue, or rather by adding up the total expenditure (consumption, investments and public spending).It is established by the INSEE for France and by the OECD for all countries. Various economic theories aim to demonstrate how technological progress and increases in workforce and investment ensure economic growth. However, there remains a question of whether these parameters are enough to lend GDP sufficient credibility.

As the idea of GDP is disputed and the vagaries of accounting begin with evaluating services, the GDP includes outright all industrial or road accidents, advertising and marketing. It ignores altruistic activity, environmental impact and the parallel economy. An initial series of amendments has been introduced to the new standard GDP, which incorporated R&D activity in 2015. In some countries, GDP takes drug markets and prostitution revenue into account.

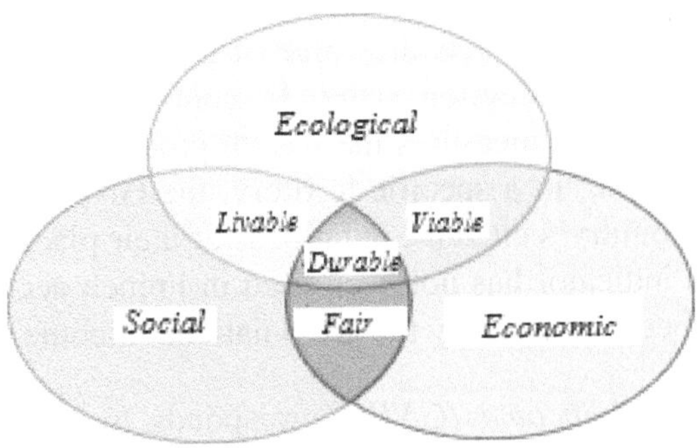

Figure A1.1. *Intersection of human activities, according to the vision for a fair and sustainable GDP*

GDP includes the market "value" of some activities whose use is not always obvious and it only partially measures the material living standards of individuals. It has been suggested that GDP should add indicators relating to employment rate, productive assets, the cumulative debt of all economic players, disability-free life expectancy, life satisfaction, income disparities, completion of higher education, carbon footprint, bird population and waste recycling rates.

The fair and sustainable GDP is one of a number of universal attempts to measure wealth (Figure A1.1). However, a consistent form of this indicator should be developed.

Moving towards a new economic index: Since 2004, the OECD has sought to create a new economic indicator. Following the "Better Life Index", a joint project between the OECD and the UN, the United Kingdom, Italy, Australia and Belgium have suggested new economic indexes intending to reflect the economic trends, quality of social life and sustainable development of all states. In June 2015, the French Economic, Social and Environmental Council (CESE – Conseil économique social et environnemental) and *France Stratégie* have joined forces to develop a set of 10 indicators compatible with GDP (http://www.lecese.fr/)

Appendix 2

Potential Harmfulness of Radio Waves

The potential dangers of waves emitted by electronic devices (particularly telephones, modems and relay masts) are the subject of renewed study in Europe and there appears to be growing evidence of their biological impact.

Electromagnetic hypersensitivity, related to exposure to waves, has been officially recognized by the Swedish Ministry of Health. If some people are affected by waves to the point of having to withdraw from all normal social and professional life, this potential resonance of human cells should bring an end to the increasing exposure of individuals, particularly children, to the low frequency waves emitted by electronic devices.

The European "Interphone" study, financed by industry and the European Commission, has demonstrated a link between excessive daily use of cell phones and the risk of developing brain cancer. "The Bioinitiative Report", published in 2007 and reprinted in 2014, analyzed over 1,800 studies establishing the effects of waves on health. According to the researchers who carried out this work, waves can have an impact, particularly on DNA, cancer, behavioral problems, autism and fertility.

French Professor Dominique Belpomme has alerted the public authorities to the risks connected to exposure to waves. It is worth recalling that all of these studies and theories are based on data from before 2010, a time when our exposure was 150 times lower than today. This is why it is recommended that earpieces are used or that the time spent with telephones pressed to ears is limited (6 min). In 2000, the average exposure to an electrical field was 0.01 volt/m^2. This figure had increased to 1.5 volts/$m^2$by 2009, even as the official recommendations advised an upper limit of 0.6 volts/m^2 (Source: *Institut pour la Protection de la Santé Naturelle* (ISPN – Natural Health Protection Institute) – Non-profit organization, Rue du vieux Marché au grain <www.institutprotectionsantenaturelle.eu>; See also [ITU 15]).

Appendix 3

ICT and Telecommunications: Sustainable Development

A joint study carried out in 2009 by IDATE and the BCG depicted the electronic consumption of the ICT industry in France and its impact on the sustainable development of the French economy. The following two main conclusions can be drawn:

1) The spread of ICT, if accompanied by judicious observance of good practices in the rest of the economy and society, can directly contribute to reducing greenhouse gas emissions by 7% between now and 2020, representing one-third of the French target according to current protocols.

2) The ICT sector's consumption in 2008 was 7.3% of French electrical consumption, or 35.3 TWh/year, and has increased more slowly than usage: 19% over 3 years against a 40% increase in the amount of user equipment and an explosion in usage. The telecommunications sector accounts for 6.7 TWh/year in this instance. Besides increasing usage, the sector's consumption of electricity has fallen by 30%. Despite the pursuit of growth in the amount of active equipment and the huge increase in their use, the voluntary moves toward energy efficiency taken by industry and public authorities could cause a reduction in the ICT sector's electricity consumption of 34.3 TWh/year by 2012 and 33.9 TWh/year by 2020. For their part, telecommunications leveled off at 8.5 TWh/year in 2012 in order to decrease to 7.6 TWh/year in 2020, once all equipment and terminals have been completely renewed.

The combined actions of industry and public authorities will therefore enable France's sustainable development to be supported through the increasing and efficient spread of ICT within the economy and society by keeping the sector's total electrical consumption stable, even as it declines slightly. For instance, optimizing peak electrical consumption could lead to a potential base of 7.9 TWh in 2020, corresponding to the total consumption of the telecommunication sector (Excerpt from the charter of the telecommunications sector's voluntary commitment to sustainable development, Paris, July 22, 2010, French Telecommunications Federation (FFT – *Fédérationfrançaise des télécoms*)).

Appendix 4

Variation in "Working Day" Telephone Traffic in 1970

Between 1950 and 1980, French telephone traffic largely reflected the country's service activity, with a busy period in the mornings close to two-thirds as large as the equivalent afternoon period and a notable break (midday to 2 p.m.) in accordance with the tradition of taking shared meals (and without cell phones or tablets within easy reach!).

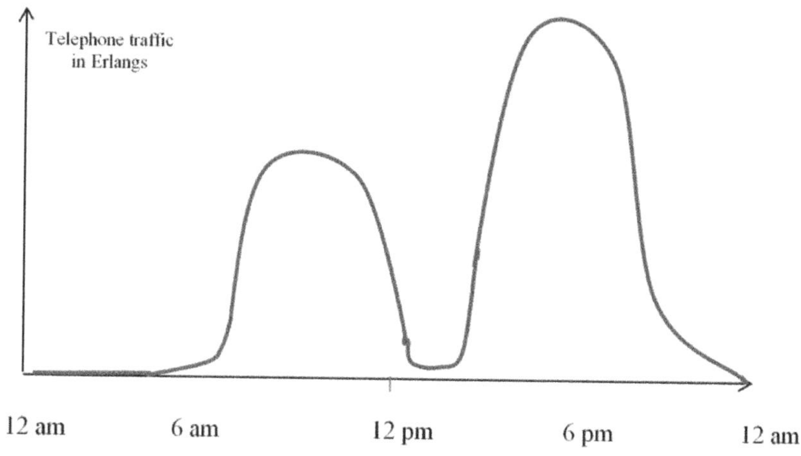

Appendix 5

Hourly Variation in National Phone Tariffs (1985)

	12–6 a.m.	6–8 a.m.	8 a.m.–12 p.m.	12–2 p.m.	2–6 p.m.	6–10 p.m.	10–12 p.m.
Germany	0.39	0.39	1	1	1	0.39	0.39
France	0.35	0.50	1	0.70	1	0.7	0.35
Italy	0.31	0.31	1	0.62	0.62	0.31	0.31
U.K.	0.40	0.40	1	0.75	0.75	0.40	0.40

The technical possibility of being able to alter tariffs throughout the day allows reductions to be offered in residential traffic. Note the differences between the midday tariffs for Italy and the United Kingdom on the one hand, and France and Germany on the other hand.

Appendix 6

Measuring the Information Society (11 October 2012) – ITU

ICT is more widely available and affordable:

– there are 6 billion mobile cellular subscriptions and 2.3 billion people (one-third of the world's population) are online;

– ICT prices dropped by 30% between 2008 and 2011;

– developing countries account for the lion's share of ICT growth and show the strongest drops in prices;

– Internet/ICT access is increasingly mobile and broadband;

– There are almost twice as many mobile broadband as fixed broadband subscriptions;

– Differences in ICT uptake, capacity and prices, between developed and developing countries (especially LCCs) persist;

– There is a need to increase broadband infrastructure and reduce prices further;

– ICT has an increasing impact on the economy;

– There is a need to stimulate investment in ICT, in particular in advanced broadband network technologies (e.g., fiber, LTE), as these are long-term investments that foster economic growth.

Appendix 7

The Development of SIM Cards

Since its creation in around 1990, the SIM card has undergone several transformations in its size (standard, nano, pico SIM), features and uses. Initially, it was simply intended to guarantee the security of electronic communications by supplying the identity of the mobile telephone user and the brand of the terminal provider. Subsequently, greater flexibility was introduced to enable users to keep the same terminal while being able to call different operators.

SIM card (2G)

The SIM card, standardized by the ETSI, was introduced with GSM mobile telephony. It brought with it the equipment and software required for its features. In GSM, the SIM card is the property of mobile operators. It has a number to identify the mobile subscriber internationally (IMSI, International Mobile Subscriber Identity) as well as the reference numbers required for subscriber authentication. Without a SIM card, the GSM user could not connect to a mobile network and would not be able to make or receive phone calls.

SIM cards enable the activities of mobile virtual network operators (MVNO),which use the resources of a network operator in the absence of their own network (around 10% of subscribers).

USIM card (3G)

For convenience, the functions of SIM cards in 3G mobile telephony (UMTS) are shared between the equipment (represented by the Universal Integrated Circuit Card (UICC)) and the software (Universal Subscriber Identity Module (USIM)). The USIM brings improvements to security, with mutual authentication, and to stronger encryption keys, as well as an improved address book. Several types of USIM card have been produced in accordance with the radio technology that has been implemented.

Mobile terminals are sometimes equipped with two USIM cards in order to ensure that calls to two different operators (for example one in France and another in the United Kingdom) can be made.

eSIM card (4G)

The eSIMcard (embedded SIM Card or Soft SIM) was created by the GSM Association at the end of 2013 with the aim of promoting machine to machine (M2M) communications technologies, particularly for the IoT and for the automotive sector, for reasons of accessibility or changes in ownership. The card is no longer installed in the mobile terminal by the network operator but by the manufacturer. It is the operator chosen by the subscriber that handles, through software on the eUICC, the necessary identification numbers (which can ask security questions later). The intelligent part of this SIM card can be linked to several network operators. The user often chooses the operator that offers them the best price for connections. The creation of this virtual universal SIM card removes the existing link between the network operator and the customer. The eSIM could be connected directly to the telephone's motherboard.

All SIM card models

In countries where mobile terminals are not subsidized by network operators (in India, Israel or Belgium, for example), all mobile telephones are unlocked. An unlocked mobile telephone is made operational, subject to identification, by providing the chosen operator with the phone's International Mobile Equipment Identity number. However, the call charges are much higher than with a subscription. At the end of 2015, the French network had 70 million SIM cards, of which 10 million were M2M.

Appendix 8

List of Recommendations from ITU-T Study Group 3

D.000	Terms and definitions for the D-series Recommendations (2010).
D.1	General principles for the lease of international (continental and intercontinental) private telecommunication circuits and networks.
D.3	Principles for the lease of analogue international circuits for private service.
D.4	Special conditions for the lease of international (continental and intercontinental) sound and television programme circuits for private service.
D.5	Costs and value of services rendered as factors in the fixing of rates.
D.7	Concept and implementation of "one-stop shopping" for international private leased telecommunication circuits.
D.8	Special conditions for the lease of international end-to-end digital circuits for private service.
D.9	Private leasing of transmitters or receivers.
D.10	General tariff principles for international public data communication services.
D.11	Special tariff principles for international packet-switched public data communication services by means of the virtual call facility.
D.12	Measurement unit for charging by volume in the international packet-switched data communication service.
D.13	Guiding principles to govern the apportionment of accounting rates in international packet-switched public data communication relations.
D.15	General charging and accounting principles for non-voice services provided by interworking between public data networks.
D.20	Special tariff principles for the international circuit-switched public data communication services.
D.21	Special tariff principles for short transaction transmissions on the international packet-switched public data networks using the fast select facility with restriction.

D.30	Implementation of reverse charging on international public data communication services.
D.35	General charging principles in the international public message handling services and associated applications.
D.36	General accounting principles applicable to message handling services and associated applications.
D.37	Accounting and settlement principles applicable to the provision of public directory services between interconnected Directory Management Domains.
D.40	General tariff principles applicable to telegrams exchanged in the international public telegram service.
D.41	Introduction of accounting rates by zones in the international public telegram service.
D.42	Accounting in the international public telegram service.
D.43	Partial and total refund of charges in the international public telegram service.
D.45	Charging and accounting principles for the international telemessage service.
D.50	International Internet connection.
D.60	Guiding principles to govern the apportionment of accounting rates in intercontinental telex relations.
D.61	Charging and accounting provisions relating to the measurement of the chargeable duration of a telex call.
D.65	General charging and accounting principles in the international telex service for multiaddress messages via store-and-forward units.
D.67	Charging and accounting in the international telex service.
D.70	General tariff principles for the international public facsimile service between public bureaux (bureaufax service).
D.71	General tariff principles for the public facsimile service between subscriber stations (telefax service).
D.73	General tariff and international accounting principles for interworking between the international bureaufax and telefax services.
D.79	Charging and accounting principles for the international videotex service.
D.80	Accounting and refunds for phototelegrams.
D.81	Accounting and refunds for private phototelegraph calls.
D.83	Rates for phototelegrams and private phototelegraph calls.
D.85	Charging for international phototelegraph calls to multiple destinations.
D.90	Charging, billing, international accounting and settlement in the maritime mobile service.
D.91	Transmission in encoded form of maritime telecommunications accounting information.
D.93	Charging and accounting in the international land mobile telephone service (provided via cellular radio systems).
D.94	Charging, billing and accounting principles for international aeronautical mobile service, and international aeronautical mobile satellite service.

D.95	Charging, billing, accounting and refunds in the data messaging land/maritime mobile satellite service.
D.96	Charging, billing, accounting and settlement principles for Global Mobile Personal Communications by Satellite (GMPCS) for the international telephone service.
D.98	Charging in international mobile roaming service.
D.99	Indicative rate for international mobile termination.
D.100	Charging for international calls in manual or semiautomatic operating.
D.103	Charging in automatic service for calls terminating on a recorded announcement stating the reason for the call not being completed.
D.104	to subscriber's station connected either to the absent subscriber's service or to a device substituting a subscriber in his absence (E.232).
D.105	Charging for calls from or to a public call office.
D.106	Introduction of reduced rates during periods of light traffic in the international telephone service.
D.110	Charging and accounting for conference calls.
D.115	Tariff principles and accounting for the International Freephone Service (IFS).
D.116	Charging and accounting principles relating to the home country direct telephone service.
D.117	Charging and accounting principles for the international premium rate service (IPRS).
D.120	Charging and accounting principles for the international telecommunication charge card service.
D.140	Accounting rate principles for the international telephone service.
D.150	New system for accounting in international telephony.
D.151	Old system for accounting in international telephony.
D.155	Guiding principles governing the apportionment of accounting rates in intercontinental telephone relations.
D.156	Network externalities.
D.160	Mode of application of the flat-rate price procedure set forth in Recommendation D.67 and Recommendation D.150 for remuneration of facilities made available to the Administrations of other countries.
D.170	Monthly telephone and telex accounts.
D.171	Adjustments and refunds in the international telephone service.
D.172	Accounting for calls circulated over international routes for which accounting rates have not been established.
D.173	Defaulting subscribers.
D.174	Conventional transmission of information necessary for billing and accounting regarding collect and credit card calls.
D.176	Transmission in encoded form of telephone reversed charge billing and accounting information.
D.177	Adjustment of charges and refunds in the international telex service.
D.178	Monthly accounts for semi-automatic telephone calls (ordinary and urgent calls, with or without special facilities).

D.180	Occasional provision of circuits for international sound- and television-programme transmissions.
D.185	General tariff and accounting principles for international one-way point-to-multipoint satellite services.
D.186	General tariff and accounting principles for international two-way multipoint telecommunication service via satellite.
D.188	General charging and accounting principles applicable to an international videoconferencing service.
D.190	Exchange of international traffic accounting data between Administrations using electronic data interchange (EDI) techniques.
D.192	Principles for charging and accounting of service telecommunications.
D.193	Special tariff principles for privilege telecommunications.
D.195	Time-scale for settlement of accounts for international telecommunication services.
D.196	Clearing of international telecommunication balances of accounts.
D.197	Notification of change of address(es) for accounting and settlement purposes.
D.201	General principles regarding call-back practices.
D.210	General charging and accounting principles for international telecommunication services provided over the Integrated Services Digital Network (ISDN).
D.211	International accounting for the use of the signal transfer point and/or signaling point for relay in Signaling System No. 7.
D.212	Charging and accounting principles for the use of Signaling System No. 7.
D.220	Charging and accounting principles to be applied to international circuit-mode demand bearer services provided over the integrated services digital network (ISDN).
D.224	Charging and accounting principles for ATM/B-ISDN.
D.225	Charging and accounting principles to be applied to frame relay data transmission service.
D.230	General charging and accounting principles for supplementary services associated with international telecommunication services provided over the Integrated Services Digital Network (ISDN).
D.231	Charging and accounting principles relating to the User-to-User Information (UUI) supplementary service.
D.232	Specific tariff and accounting principles applicable to ISDN supplementary services.
D.233	Charging and accounting principles to be applied to the reversed charge supplementary service.
D.240	Charging and accounting principles for teleservices supported by the ISDN.
D.250	General charging and accounting principles for non-voice services provided by interworking between the ISDN and existing public data networks.

D.251	General charging and accounting principles for the basic telephone service provided over the ISDN or by interconnection between the ISDN and the public switched telephone network.
D.260	Charging and accounting capabilities to be applied on the ISDN.
D.271	Charging and accounting principles for NGN (2016).
D.280	Principles for charging and billing, accounting and reimbursements for universal personal telecommunication.
D.285	Guiding principles for charging and accounting for intelligent network supported services.
D.286	Charging and accounting principles for the global virtual network service.
D.300R	Determination of accounting rate shares in telephone relations between countries in Europe and the Mediterranean Basin.
D.301R	Determination of accounting rate shares and collection charges in telex relations between countries in Europe and the Mediterranean Basin.
D.302R	Determination of the accounting rate shares and collection charges for the international public telegram service applicable to telegrams exchanged between countries in Europe and the Mediterranean Basin.
D.303R	Determination of accounting rate shares and collection charges applicable by countries in Europe and the Mediterranean Basin to the occasional provision of circuits for sound- and television programme transmissions.
D.307R	Remuneration of digital systems and channels used in telecommunication relations between the countries of Europe and the Mediterranean Basin.
D.310R	Determination of rentals for the lease of international programme (sound and television) circuits and associated control circuits for private service in relations between countries in Europe and the Mediterranean Basin.
D.400R	Accounting rates applicable to direct traffic relations in voice telephony between countries in Latin America and the Caribbean.
D.500R	Accounting rates applicable to telephone relations between countries in Asia and Oceania.
D.501R	Accounting rates applicable to telex relations between countries in Asia and Oceania.
D.600R	Cost methodology for the regional tariff group for Africa applicable to the international automatic telephone service.
D.601R	Determination of accounting rate shares and collection charges in telex relations between countries in Africa.
D.602R	Application of the “sender pays transit” principle in transit relation.
D.603R	Minimizing collection charges on inter-African calls.
D.604R	The last international transit center pays the traffic.
D.606R	Preferential rates in telecommunication relations between countries in Africa.
D.Sup1	Cost and tariff study method.
D.Sup2	Method for carrying out a cost price study by regional tariff groups.
D.Sup3	Handbook on the methodology for determining costs and establishing national tariffs.

Appendix 9

The Cost of Capital

The cost of capital is an economic indicator that shows the annual rate of return expected by stockholders and creditors in exchange for their investment. It takes different risk factors into account:

– the inflationary risks linked to currency depreciation;

– the country-specific risk;

– the systematic risk;

– the return on investment.

In the new economic environment, the cost of capital remains the basis for value calculations in methods of assessment and cost models.

Appendix 10

NGN and the Internet in Figures

In 30 years, the Internet has conquered over half the globe and the NGN standards which connect fixed-line networks to mobile networks have shown the consistency of the plans initialized several decades earlier. These positive results demonstrate the merits of the proposals made internationally on the political and commercial levels.

Connections

New generation networks (NGNs) ensure the inter functionality of digital circuit-switched telephone networks with digital mobile telephone networks and the Internet. Internet access through a digital circuit-switched telephone network requires the use of compatible dial-up modems or connection through a suitable digital link (ADSL, VDSL2). No mobile telephone terminals consistently provide access to all multimedia applications available on the Internet. A broadband Internet connection is a guarantee of good quality of service for multimedia applications.

The Internet's place in the world at the end of 2015

At the end of 2015, the number of subscribers in mobile cellular telephony surpassed 7 billion (usage rate of 97%) and 9.5 billion mobile subscriptions are expected by the end of 2020. Between 2000 and 2015, global Internet penetration rose from 6.5 to 43%, a sevenfold increase. Fixed broadband technology is spreading at a slower rate and reached 11% penetration by the end of 2015. Mobile broadband is the most dynamic part of the market worldwide and had a penetration rate of 47% at the end of 2015, 12 times the 2007 figure. The proportion of households with Internet access has risen from 18% in 2005 to 46% in 2015.

The number of Internet users worldwide has more than tripled in 10 years, increasing from a billion in 2005 to 3.2 billion at the end of 2015, of whom 2 billion have access through a mobile telephone. However, “the effects of this technology on global productivity and improving opportunities for the poor and the middle class have not lived up to expectations”, according to a report by the World Bank, which highlights some negative effects on the labor market.

Almost 4.2 billion people, close to 60% of the world’s population, still do not have access to the Internet. Broadband Internet only serves 1.1 billion people, or less than 15% of the worldwide population. India has the highest number of people without Internet connections (1.1 billion), followed by China (755 million) and Indonesia (213 million).

Internet penetration is assessed according to the “proportion of the population, which has the opportunity to access this network using a computer, mobile telephone, personal digital assistant, games consoles, television sets” and so on. In 2014, the statistics prepared by the World Bank show that while the level of Internet penetration per 100 inhabitants is very high in Finland and Denmark (96%), it remains very low (between 1.38 and 4.8%) in sub-Saharan Africa (Benin, Ethiopia, Democratic Republic of the Congo, Tanzania). At the end of 2015, China recorded 600 million Internet users (httpp://data.un.org/).

In OECD countries, eight adults in 10 are connected to the Internet. Approximately half of Internet users visit social networks and purchase goods and services online. Currently, 41% of Internet users connect through a mobile telephone or smartphone. In terms of usage, six Internet users in ten visit social networks and 50% purchase goods and services online. Merely one-third of users have recourse to electronic administration services and only one in 10 makes medical appointments on the Internet. Furthermore, in 2012, small businesses were almost all connected in Canada, Denmark, Finland, France, Iceland, South Korea and Switzerland, and only 21% of them offered products for sale online.

In Europe

In 2014, almost 75% of subscribers in the European Union received broadband. The average speed increased from 30 Mbps in 2013 to 38 Mbps in 2014. The European Commission is proposing to increase the number of broadband connections (at least on a par with 30 Mbps) in rural areas of Europe through the proliferation of long-term loans. The European Union performs better than the United States in terms of Internet access speeds. ADSL technologies provide speeds of 8.27 Mbps in Europe and 7.67 Mbps in the United States, with cable delivering 66.57 Mbps in France and 25.48 Mbps in the United States and fiber 53.09 Mbps in France and 41.35 Mbps in the United States.

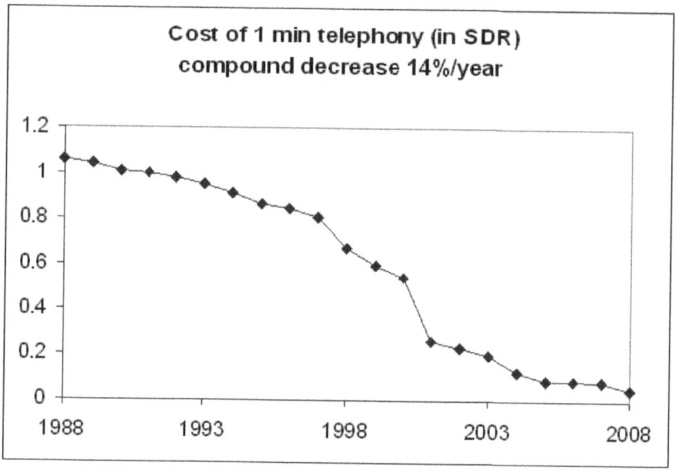

Figure A10.1. *Decrease in telephony costs (in SDR) between 1988 and 2008 (Source ITU-T)*

Almost 95% of residents and businesses in the English region of Cornwall will soon have access to a very high speed fixed-line Internet connection. The major advantages that this technology brings are related to being able to put videos and other content online more quickly, the opportunity to work from home, making conference calls without interruption or even international communications using VoIP. The British want to make Cornwall a European Silicon Valley, an ambitious program for this region of 500,000 inhabitants, which should welcome a cluster linked to the universities of Falmouth and Plymouth.

Broadband prices went down by 12% in the European Union between 2012 and 2015, and by almost 20% for speeds ranging from 30 to 100 Mbps. The disparities between European countries can be large. Over 99.4% of European homes had access to at least one fixed or mobile broadband technology in 2014, according to a study carried out by the European Commission.

In France

According to Arcep's annual survey, in 2015, 58% of the French population had a smartphone and 8% of French people still had no mobile telephone. In spite of this, the use of mobile telephone equipment (92%) still surpasses fixed-line equipment (89%). In 1 year, the average consumption by 3G and 4G subscribers almost doubled to reach 1.1 Gbps. Because of 4G, Orange's mobile subscriber base passed the 24 million mark (+10%), in which the number of customers has reached 8 million. With regard to fixed line, very high speed broadband is making progress: it increased from 9 to 14% of connections in homes. Fiber-optic connection, on the other hand, has barely taken off (4% against 3% in 2014). The French are coming around to social networks a little more each year: in 2015, 62% of them were registered on Facebook, Twitter or LinkedIn.

Orange set a target of 12 million connectable French households in 2018 and 20 million by 2022. At the end of 2015, Orange had a total of almost 1.9 million customers in fiber-optics and almost 13 million homes connectable by fiber-optics. In the French business market,

Orange is continuing with development of Fiber To The Office (FTTO) with over 31,000 connected customer sites to date.

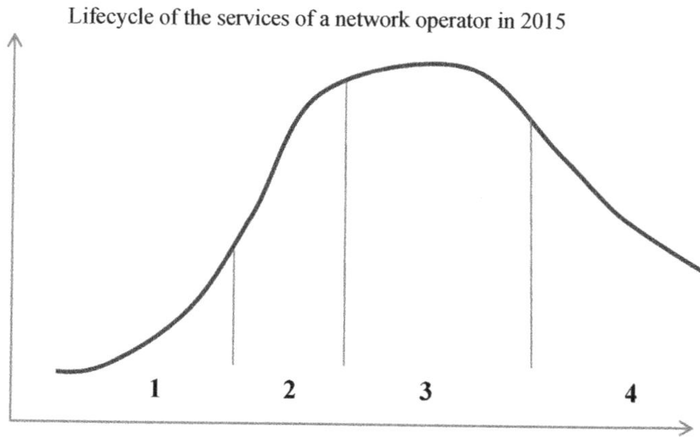

Figure A10.2. *Lifecycle of services in 2015: 1) Introduction – Internet of hings – Mobile financial transactions – 5G; 2) Developments – Broadband – Fiber-optic; 3) Maturity – DSL – 3 and 4G; 4) Decline – RTC – Gigabit Ethernet.*

Appendix 11

Standardization of Mobile Financial Services (Extract – ITU-T, February 2016)

We live in a global economy that is increasingly dependent on mobile telecommunications technology and digital financial services. Mobile communications offer financial and banking operations that are profitable both nationally and internationally, particularly for the world's low-income population.

These services concern mobile money, mobile payment and mobile banking services with increased security. Features for transfers, payments and savings are not available for the poorest people anywhere in the world. As a result, these financial services are essential to the benefit of over two billion people living on less than three dollars per day.

For these reasons, the ITU-T needs to help define a "mobile financial services ecosystem" and the technologies required for proper interoperability connected to a high level of security, within a regulatory framework suited to users.

Appendix 12

The Internet's Flaws (Vint Cerf)

The Internet remains an unfinished and evolving work. We are reminded daily of opportunities for improvement. Recent revelations concerning surveillance of Internet traffic suggest that we should take a number of steps to protect user privacy and confidentiality. The regular use of end-to-end cryptography at multiple layers in the protocol architecture seems an obvious improvement.

Recently, reports have surfaced of so-called "buffer-float" problems caused by the improper sizing of buffers at firewalls or routers that lie at a discontinuity in local and Internet access bandwidth. This creates an increasing delay and misinterpretation of resulting measured round-trip time [CER 14].

Bibliography

[ACE 13] ACEMOGLU D., ROBINSON J.A., *Why Nations Fail: The Origins of Power, Prosperity and Poverty*, Crown Business, Danvers, MA, 2013.

[AND 06] ANDERSON C., *The Long Tail: Why the Future of Business is Selling Less of More*, Hyperion, New York, 2006.

[ARC 12a] ARCEP, L'Actualité de l'Arcep – Les chiffres des communications électroniques et postales en France, Paris, 2012.

[ARC 12b] ARCEP, Les Cahiers de l'ARCEP, Paris, June 2012, available at http://www.slate.fr/.

[ART 15] ARTUS P., VIRARD M.-P., *Croissance Zéro*, Fayard, 2015.

[BAI 15] BAIN & COMPANY, *Generation Hashtag – The Rise of the Digital Native*, Bain & Company, 2015.

[BAN 12] BANERJEE A.V., DUFLO E., *Repenser la pauvreté*, Le Seuil, Paris, 2012.

[BAR 14] BARICCO A., *Les barbares. Essai sur la mutation*, Gallimard, 2014.

[BAR 15] BARDI U., *Le Grand Pillage – Comment nous épuisons les ressources de la planète*, Les petits matins, 2015.

[BAT 78] BATRA R., *The Downfall of Capitalism and Communism: A New Study of History*, Palgrave Macmillan, New York, 1978.

[BAT 84] BATTU D., ROSE J.B., Telecommunication Services for the Transfer of Information and Data: A Case Study in Indonesia, UNESCO, Paris, 1984.

[BAT 02] BATTU D., *Télécommunications : Principes, infrastructures et services*, Dunod, Paris, 2002.

[BAT 07] BATRA R., *The New Golden Age: The Coming Revolution against Political Corruption and Economic Chaos*, Editions Palgrave Macmillan Trade, 2007.

[BAT 13] BATRA R., Wireless Network Pricing – Synthesis Lectures on Communication Networks, June 2013.

[BAT 14] BATTU D., *New Telecom Networks: Enterprises and Security*, ISTE, London and Wiley, New York, 2014.

[BAT 15] BATRA R., *End Unemployment Now: How to Eliminate Poverty, Debt and Joblessness Despite Congress*, St. Martin's Press, 2015.

[BAU 15] BAUWENS M., *Sauver le monde*, Editions Les Liens qui Libèrent, Paris, 2015.

[BER 84] BERTHO C., "Les réseaux téléphoniques de Paris – 1879–1927", *Réseaux*, no. 4, 1984.

[BLA 11] BLACKMAN C., SRIVASTAVA L., *Telecommunications Regulation Handbook*, Tenth Anniversary Edition, ITU, Geneva, 2011.

[BOU 15] BOUBA-OLGA O., GROSSETTI O., "La métropolisation, horizon indépassable de la croissance économique ?", *Revue de l'OFCE*, no. 143, 2015.

[BOU 16] BOURREAU M., *Régulation des communications électroniques*, Lavoisier, Paris 2016.

[BUR 07] BURROUGH B., *Barbarians at the Gate: The Fall of Rjr Nabisco*, Collins Business, 2007.

[CAG 16] CAGE J., *Saving the Media: Capitalism, Crowdfunding and Democracy*, Harvard University Press, 2016.

[CER 14] CERF V., "Unfinished business", *IEEE Internet Computing*, Piscataway, NJ USA, 2014.

[CHR 33] CHRISTALLER W., Central Places in Southern Germany, University of Jena, 1933.

[COH 15] COHEN D., *Le Monde est clos et le désir infini*, Albin Michel, Paris, 2015.

[COL 12] COLIN N., VERDIER H., *L'Âge de la multitude*, Odile Jacob, Paris, 2012.

[COL 15] COLINS N., *La Richesse des Nations après la Révolution Numérique*, Terra Nova, Paris, 2015.

[COU 14] COUPPEY-SOUBEYRAN J., NIJDAM C., *Parlons banque en 30 questions*, La Documentation française, Paris, 2014.

[DAG 13] DAGNAUD M., *Génération Y : Les jeunes et les réseaux sociaux, de la dérision à la subversion*, Presses de Sciences-Po, Paris, 2013.

[DEN 66] DENIS H., *Histoire De La Pensée Économique*, Editions PUF – Thémis, Paris, 1966.

[DES 10] DESBRUYÈRES D., *Le Trésor des abysses*, Editions Quae, 2010.

[DUP 12] DUPUY J.-P., *L'Avenir de l'économie : Sortir de l'écomystification*, Flammarion, Paris, 2012.

[ECO 15] ECONOMIC ANALYSIS COUNCIL, Paris, February 2015.

[ESC 15] ESCANDE P., CASSINI S., *Bienvenue dans le capitalisme 3.0*, Albin Michel, Paris, 2015.

[FIE 14] FIEVET C., *Comprendre Bitcoin et les crypto-monnaies alternatives*, Librinova, Paris, 2014.

[GIL 15] GILLI F., "Les territoires de l'innovation... ne sont pas ceux qu'on croit", *Alternatives Economiques*, no. 068, 2015.

[GRA 47] GRAVIER J.-F., *Paris et le désert français,* Flammarion, Paris, 1947.

[GUI 12] GUILHON B., *Les paradoxes de l'économie du savoir*, Hermes-Lavoisier, Paris, 2012.

[HOE 16] HOEFFLINGER B., *CHIPS 2020* vol. *2 – New Vistas in Nanoelectronics*, Springer, New York, 2016.

[HOU 91] HOUELLEBECQ M., *H. P. Lovecraft. Contre le monde, contre la vie*, Editions du Rocher, Paris, 1991.

[HUA 13] HUANG J., Wireless Network Pricing – Synthesis Lectures on Communication Networks, The Chinese University of Hong Kong and Lin Gao, 2013.

[ICT 15] ICT Facts and Figures 2015, ITU, available at http://www.itu.int/en/ITU-D/Statistics/Documents/facts/ICTFactsFigures2015.pdf, Geneva, 2015.

[ID 97] THE INNOVATOR'S DILEMMA, *When New Technologies Cause Great Firms to Fail*, Harper Business, New York, 1997.

[IEE 15] IEEE COMMUNICATIONS, Underwater Wireless Communications and Networks, 2015.

[IEE 15] IEEE Spectrum, The Multiple Lives of Moore's Law, April 2015.

[IFR 12] IFRI, "Internet, outil de puissance", *Politique Etrangère*, vol. 77, no. 2, 2012.

[ITU 11] ITU-T, Security in Telecommunications and Information Technology. An overview of issues and the deployment of existing ITU-T Recommendations for secure telecommunications, Geneva, 2011.

[ITU 14] ITU, Measuring the information society, Report, Geneva, 2014.

[ITU 15] ITU-T Study Group 5, Activities on Human Exposure to Electromagnetic Fields (EMFs) Due to Radio Systems and Mobile Equipment, Geneva, 2015.

[ITU 15a] ITU, Measuring the information society report 2015. Geneva.

[ITU 15b] ITU-T Study Group 5, Activities on human exposure to electromagnetic fields (EMFs) due to radio systems and mobile equipment, ITU-T Study Group 5, Geneva, 2015.

[JEA 15] JEANNOT G., COUTARD O., *Revenir au service public ?*, La Documentation française, Paris, 2015.

[JIP 63] JIPP A., "Wealth of nations and telephone density", *ITU Telecommunication Journal*, Geneva, 1963.

[JUR 60] JURAN J., "Pareto, Lorenz, Cournot, Bernouli, Juran and others", *Industrial Quality-Control*, vol. 17, no. 4, McGraw-Hill, 1960.

[KEL 05] KELLY T., *Acacia Atlas*, IDRC, Ottawa, October 2005.

[KEN 08] KENNEDY M, LIETAER B., *Monnaies régionales*, Editions Mayer Charles Leopold, Paris, 2008.

[KIR 11] KIRKPATRICK D., *La révolution facebook*, JC Lattes, Paris, 2011.

[KON 26] KONDRATIEV N., *The Major Cycles of the Conjecture*, Pickering & Chatto, London, 1926.

[LAB 73] LABORDE P., "Le télex, un aspect mal connu de la vie de relations", *Annales de Géographie*, Paris, vol. 82, no. 450, pp. 193–207, 1973.

[LAN 10] LAND REGISTER, French Forestry Portal, 2010.

[LAU 16] LAURENT E., *Nos mythologies économiques*, Les liens qui libèrent, Paris, 2016.

[LEM 02] LEMESLE R.M., *L'économie des télécommunications en Afrique*, Karthala, 2002.

[LEW 14] LEWIS M., *Flash Boys*, W.W. Norton & Company, 2014.

[LÖS 54] LÖSCH A., *The Economics of Location*, Yale University Press, 1954.

[MOO 65] MOORE G.E., "Cramming more components onto integrated circuits", *Electronics*, vol. 38, pp.114–117, 1965.

[MUL 14] MULAY A., "Mass capitalism: a blueprint for economic revival", *Sustaining Moore's Law: Uncertainty Leading to Certainty of IoT Revolution*, Book Publishers Network, 2014.

[NGU 00] NGUYEN G.D., PHAN D., "The economics of telecommunications and the Internet", *Economica*, 2000.

[OEC 92] OECD, Communications Outlook Paris, 1992.

[OEC 09] OECD, Communications Outlook, Paris 2009.

[OEC 11] OECD, Communications Outlook, Paris, 2011.

[OST 10] OSTERWALDER A., PIGNEUR Y., *Business Model Generation – A Handbook for Visionaries, Game Changers, and Challengers*, John Wiley & Sons, 2010.

[PAU 78] PAUTRAT C., GUERARD A., FAUVET A., "Analyse de l'évolution des régions et départements français au travers de leur équipment téléphonique", *Les Annales des Télécommunications*, Paris, vol. 7, no. 8, pp. 2–18, 1978.

[PIK 13] PIKETTY T., "Capital in the Twenty-First Century", *The Economics of Inequality*, 2013.

[PLI 13] PLIHON D., *La monnaie et ses mécanismes*, La Découverte, Paris, 2013.

[POL 83] POLANYI K., *La Grande Transformation. Aux origines politiques et économiques de notre temps*, Gallimard, Paris, 1983.

[REY 15] REY H., PASSARI E., "Financial flows and the international monetary system", *Economic Journal*, vol. 125, no. 584, pp. 675–698, Royal Economic Society, May 2015

[RIF 14] RIFKIN J., *La nouvelle société du coût marginal zéro*, Les liens qui libèrent, 2014.

[ROD 07] RODRIK D., *One Economics, Many Recipes*, Princeton University Press, 2007.

[SCH 74] SCHUMPETER J., *Capitalisme, socialisme et démocratie*, Petite Bibliothèque Payot, 1974.

[SER 99] SERVET J.-M., *Une économie sans argent. Les Systèmes d'Echange Local*, Le Seuil, 1999.

[SOL 00] SOLOW R.M., *Growth Theory: An Exposition*, Oxford University Press, 2000.

[SOW 06] SOWELL T., *On Classical Economics*, Yale University Press, 2006.

[STA 15] STAROSIELSKI N., *The Undersea Network*, Duke University Press, 2015.

[TRE 15] TRENDS IN TELECOMMUNICATION REFORM, "Getting ready for the digital economy", available at http://www.itu.int/en/publications/Documents/Trends2015-short-version_pass e374681.pdf, 2015.

[UK 15] UK HOUSE OF LORDS, Make or break: The UK's digital future, Digital Skills Committee report, May 2015.

[UNI 02] UNIVERSITÉ DE TOUS LES SAVOIRS, "Les modèles géométriques de réseau urbain", *Geography and Demography*, Odile Jacob, 2002.

[VAN 15] VANOT J.-P., "The evolution of the telecommunications sector", *REE Review,* No. 4, 2015.

[WNE XX] HUANG J., "Wireless network economics and games", *Network Communications & Economics Lab*, The Chinese University of Hong Kong.

Glossary

3G: High-speed radio technology (between 1.5 and 3 Mbps).

4G: Radio technology with speeds above 100 Mbps, even reaching 1 Gbps.

5G: Radio technology with speeds exceeding several Gbps.

Accounting rate: The rate agreed between administrations in a given relationship that is used for the establishment of international accounts (Rec. D.000).

ADSL (asymmetric digital subscriber line): Allows a telephone line subscriber to receive all Internet-related services at a data rate 10 times higher than analog modems (ADSL reaching 7 Mbps and ADSL2+ reaching 20 Mbps).

ATM: Asynchronous transfer mode.

B2B and C2C: Business-to-Business, and Consumer-to-Consumer (or Customer-to-Customer) relationships.

Bluetooth: Close range radio technology.

Broadband: Regulatory authorities currently consider subscriber services between 30 and 100 Mbps to be broadband. Superfast broadband refers to connections above 100 Mbps.

CDN: Content delivery network (multimedia).

Collection charge: Charge established and collected by a network operator from its customers for the use of an international telecommunication service.

Cybersecurity: All means and measures required to guarantee Internet security. "The state sought by a computer system to allow it to withstand all occurrences in cyberspace".

Darknet: System of anonymized and encrypted networks, including the Tor, I2P and FreeNet networks, which host sites carrying an unreferenced DNS address suffix.

Direct transit country: Transit country through which traffic is routed on direct circuits, namely circuits provided for the exclusive use of other countries.

DTT: Digital terrestrial television. Television service with various programs that is conveyed through regional or satellite radio transmitters.

DOCSIS: Data Over Cable Service Interface Specification – standard for connecting the subscriber to 10/1 Gbps broadband for multimedia services.

Flow: A set of IP packets passing an observation point in the network during a certain time interval [ITU-T Y.2233].

GAFA: Abbreviation that indicates the Internet's major players: Google, Apple, Facebook and Amazon.

GbE: Gigabit Ethernet – standards for connecting Ethernet subscribers at speeds between 1 Gbps and 1,000 Gbps, largely on fiber-optics.

HCF: Hybrid Coaxial Fiber: Subscriber connection made through fiber-optics and coaxial cable.

HSBB: Very high-speed broadband, both in the downlink direction (from the network to the subscriber) and kbps in the uplink direction (from the subscriber to the network).

International circuit: Circuit between international centers located in different countries (Rec. D.000).

International mobile roaming (IMR): A service that allows subscribers to continue to use their mobile telephones and their telephone number to access voice and short messaging services abroad through the mobile network of the country visited.

IPv4/IPv6 : Addressing of Internet terminals.

IPX: Internetwork Packet eXchange: Provides a platform for connecting IP services between mobile and fixed networks and service providers.

IXP – Internet eXchange Point – GIX (Global Internet eXchange): Physical access point shared between Internet service providers and networks, use of which is determined by peering agreements.

IXP: An Internet eXchange Point (IXP) is a physical network access point through which major network providers connect their networks and exchange traffic. The primary focus of an exchange point is to facilitate network interconnection through an exchange access point instead of third-party networks and allows the local circulation of traffic routed in the region. It allows savings to be made in internetwork throughput capacity, which reduces the overall costs of Internet connectivity.

MFS – Mobile financial services: MFS allows mobile telephones to be used to carry out financial transactions, such as:

– depositing funds, transactions and payments;

– money transfers (including P2P, C2B, B2B);

– the mobile electronic wallet (e-wallet);

– government-to-person (G2P) payments, which include social transfers as well as salary and pension payments;

– account information services.

These services can be offered by an MNO, an independent payment platform with banking connections, or by a bank itself. All of these actors are considered “mobile payment operators” (MPO).

MVNO: Mobile virtual network operator.

National extension: The part of the connection that runs from the national section of an international center to the subscriber.

NGN: New generation network.

Non-discrimination: Treating a service or service provider no less favorably than other identical services or service providers in similar conditions.

NRA: National Regulatory Authorities.

On-Net: Intranetwork call or message. The term “On-Net” is used when a call or message originates on the initial operator’s network and finishes at another mobile telephone number close to that operator.

Off-Net: Call or message outside networks. “Off-Net” is used when the call or message comes from a different network, when roaming, for example, or through a domestic network when the call or message is directed toward a number belonging to a different network provider.

OTT (Over The Top): “High-level actors” not represented at the ITU (International Telecommunication Union) which offer original applications, either for a price or for free, and with or without Internet advertising. An OTT service is one that bypasses the distribution channels of traditional network operators on the Internet. This circumvention can potentially place these services and applications in competition or see them stand in for traditional telecommunications or broadcasting services, such as voice telephony, SMS, video and voice calls, video streaming and downloading, all of which may or may not be connected to advertisements. The providers of these OTT applications are also called "OTT". Online Internet services that are not in competition with traditional telecommunications or broadcasting services are not included in this field.

P2P: Peer-to-peer.

PCM: Pulse code modulation.

PLC: Power-line carrier technology.

PTSN: Public switched telephone network. This uses the same technology as analog and digital circuit-switching.

RCS: Rich communications services – platform offering the distribution of voice communications and SMS messaging data through instant messaging, live video or file sharing.

Router: A piece of network equipment that, located between several servers, allows users to access network resources.

SDH: Synchronous digital hierarchy. A broadband digital transmission system.

Session: A temporary telecommunications relationship among a group of objects in the service stratum that is assigned to collectively fulfill a task for a period of time [ITU-T Y.2091].

Settlement rate: The part of the accounting rate corresponding to the means available to each country; this share is fixed by agreement between administrations.

SMS: Short message service – A text message (or "text") limited to 140 or 160 characters and exchanged between mobile telephones or sent from a computer connected to the Internet.

TCP: Transmission control protocol.

Terminal rate: The part of the accounting rate which returns to the end administration.

Termination charge: Charge fixed by the destination network operator for incoming traffic regardless of origin.

ToIP: Telephony over IP.

Transit rate: The part of the accounting rate that returns to an intermediary administration whose facilities or circuits are borrowed to route traffic between two end countries.

Uberization: (From "Uber", the major Californian platform created for car hire services with drivers). Creation of a part-time business to provide services through an intermediary computer platform connected to the Internet.

UDP: User datagram protocol.

UHF: Ultra high frequency. Radio spectrum band ranging from 300 to 3,000 MHz.

Unbundling, partial or total: Unbundling allows Internet service providers to use old France Telecom infrastructure to develop their offers. In partial unbundling, the fixed telephone line continues to be managed by the traditional operator. In total unbundling, all services (Internet, telephony) are managed by the Internet service provider (ISP).

VDSL: Very-high-bit-rate DSL – Broadband xDSL technology on metallic subscriber pairs (from 1 to 55 Mbps).

Wi-Fi: Radio technology that allows several mobile data terminals to be linked over an Internet network connection according to IEEE802.11 standards. The routers sold by network operators and ISPs have this technology.

Index

Other titles from

iSTE

in

Networks and Telecommunications

2016

AL AGHA Khaldoun, PUJOLLE Guy, ALI-YAHIYA Tara
Mobile and Wireless Networks (Advanced Network Set – Volume 2)

BENSLAMA Malek, BATATIA Hadj, MESSAI Abderraouf
Transitions from Digital Communications to Quantum Communications: Concepts and Prospects

CAMARA Daniel, NIKAEIN Navid (Eds.)
Wireless Public Safety Networks 2: A Systemic Approach

CHIASSERINI Carla Fabiana, GRIBAUDO Marco, MANINI Daniele
Analytical Modeling of Wireless Communication Systems (Stochastic Models in Computer Science and Telecommunication Networks Set – Volume 1)

EL FALLAH SEGHROUCHNI Amal, ISHIKAWA Fuyuki, HÉRAULT Laurent, TOKUDA Hideyuki
Enablers for Smart Cities

PEREZ André
VoLTE and ViLTE

RAMOS Angel, LAZARO Antonio, GIRBAU David, VILLARINO Ramon
RFID and Wireless Sensors using Ultra-Wideband Technology

SENOUCI Mustapha Reda, MELLOUK Abdelhamid
Deploying Wireless Sensor Networks

2015

BENSLAMA Malek, KIAMOUCHE Wassila, BATATIA Hadj
Connections Management Strategies in Satellite Cellular Networks

BENSLAMA Malek, BATATIA Hadj, BOUCENNA Mohamed Lamine
Ad Hoc Networks Telecommunications and Game Theory

BERTHOU Pascal, BAUDOIN Cédric, GAYRAUD Thierry, GINESTE Matthieu
Satellite and Terrestrial Hybrid Networks

CÂMARA Daniel
Bio-inspired Networking

CÂMARA Daniel, NIKAEIN Navid (Eds.)
Wireless Public Safety Networks 1: Overview and Challenges

CUADRA-SANCHEZ Antonio, ARACIL Javier
Traffic Anomaly Detection

JARRY Pierre, BENEAT Jacques N.
Digital Communications

LE RUYET Didier, PISCHELLA Mylène
Digital Communications 1: Source and Channel Coding

PEREZ André
LTE and LTE Advanced: 4G Network Radio Interface

PISCHELLA Mylène, LE RUYET Didier
Digital Communications 2: Digital Modulations

PUJOLLE Guy
Software Networks (Advanced Network Set – Volume 1)

2014

ANJUM Bushra, PERROS Harry
Bandwidth Allocation for Video under Quality of Service Constraints

BATTU Daniel
New Telecom Networks: Enterprises and Security

BEN MAHMOUD Mohamed Slim, GUERBER Christophe, LARRIEU Nicolas, PIROVANO Alain, RADZIK José
Aeronautical Air–Ground Data Link Communications

BITAM Salim, MELLOUK Abdelhamid
Bio-inspired Routing Protocols for Vehicular Ad-Hoc Networks

CAMPISTA Miguel Elias Mitre, RUBINSTEIN Marcelo Gonçalves
Advanced Routing Protocols for Wireless Networks

CHETTO Maryline
Real-time Systems Scheduling 1: Fundamentals
Real-time Systems Scheduling 2: Focuses

EXPOSITO Ernesto, DIOP Codé
Smart SOA Platforms in Cloud Computing Architectures

MELLOUK Abdelhamid, CUADRA-SANCHEZ Antonio
Quality of Experience Engineering for Customer Added Value Services

OTEAFY Sharief M.A., HASSANEIN Hossam S.
Dynamic Wireless Sensor Networks

PEREZ André
Network Security

PERRET Etienne
Radio Frequency Identification and Sensors: From RFID to Chipless RFID

REMY Jean-Gabriel, LETAMENDIA Charlotte
LTE Standards
LTE Services

TANWIR Savera, PERROS Harry
VBR Video Traffic Models

VAN METER Rodney
Quantum Networking

XIONG Kaiqi
Resource Optimization and Security for Cloud Services

2013

ASSING Dominique, CALÉ Stéphane
Mobile Access Safety: Beyond BYOD

BEN MAHMOUD Mohamed Slim, LARRIEU Nicolas, PIROVANO Alain
Risk Propagation Assessment for Network Security: Application to Airport Communication Network Design

BEYLOT André-Luc, LABIOD Houda
Vehicular Networks: Models and Algorithms

BRITO Gabriel M., VELLOSO Pedro Braconnot, MORAES Igor M.
Information-Centric Networks: A New Paradigm for the Internet

BERTIN Emmanuel, CRESPI Noël
Architecture and Governance for Communication Services

DEUFF Dominique, COSQUER Mathilde
User-Centered Agile Method

DUARTE Otto Carlos, PUJOLLE Guy
Virtual Networks: Pluralistic Approach for the Next Generation of Internet

FOWLER Scott A., MELLOUK Abdelhamid, YAMADA Naomi
LTE-Advanced DRX Mechanism for Power Saving

JOBERT Sébastien *et al.*
Synchronous Ethernet and IEEE 1588 in Telecoms: Next Generation Synchronization Networks

MELLOUK Abdelhamid, HOCEINI Said, TRAN Hai Anh
Quality-of-Experience for Multimedia: Application to Content Delivery Network Architecture

NAIT-SIDI-MOH Ahmed, BAKHOUYA Mohamed, GABER Jaafar, WACK Maxime
Geopositioning and Mobility

PEREZ André
Voice over LTE: EPS and IMS Networks

2012

AL AGHA Khaldoun
Network Coding

BOUCHET Olivier
Wireless Optical Communications

DECREUSEFOND Laurent, MOYAL Pascal
Stochastic Modeling and Analysis of Telecoms Networks

DUFOUR Jean-Yves
Intelligent Video Surveillance Systems

EXPOSITO Ernesto
Advanced Transport Protocols: Designing the Next Generation

JUMIRA Oswald, ZEADALLY Sherali
Energy Efficiency in Wireless Networks

KRIEF Francine
Green Networking

PEREZ André
Mobile Networks Architecture

2011

BONALD Thomas, FEUILLET Mathieu
Network Performance Analysis

CARBOU Romain, DIAZ Michel, EXPOSITO Ernesto, ROMAN Rodrigo
Digital Home Networking

CHABANNE Hervé, URIEN Pascal, SUSINI Jean-Ferdinand
RFID and the Internet of Things

GARDUNO David, DIAZ Michel
Communicating Systems with UML 2: Modeling and Analysis of Network Protocols

LAHEURTE Jean-Marc
Compact Antennas for Wireless Communications and Terminals: Theory and Design

RÉMY Jean-Gabriel, LETAMENDIA Charlotte
Home Area Networks and IPTV

PALICOT Jacques
Radio Engineering: From Software Radio to Cognitive Radio

PEREZ André
IP, Ethernet and MPLS Networks: Resource and Fault Management

TOUTAIN Laurent, MINABURO Ana
Local Networks and the Internet: From Protocols to Interconnection

2010

CHAOUCHI Hakima
The Internet of Things

FRIKHA Mounir
Ad Hoc Networks: Routing, QoS and Optimization

KRIEF Francine
Communicating Embedded Systems / Network Applications

2009

CHAOUCHI Hakima, MAKNAVICIUS Maryline
Wireless and Mobile Network Security

VIVIER Emmanuelle
Radio Resources Management in WiMAX

2008

CHADUC Jean-Marc, POGOREL Gérard
The Radio Spectrum

GAÏTI Dominique
Autonomic Networks

LABIOD Houda
Wireless Ad Hoc and Sensor Networks

LECOY Pierre
Fiber-optic Communications

MELLOUK Abdelhamid
End-to-End Quality of Service Engineering in Next Generation Heterogeneous Networks

PAGANI Pascal *et al.*
Ultra-wideband Radio Propagation Channel

2007

BENSLIMANE Abderrahim
Multimedia Multicast on the Internet

PUJOLLE Guy
Management, Control and Evolution of IP Networks

SANCHEZ Javier, THIOUNE Mamadou
UMTS

VIVIER Guillaume
Reconfigurable Mobile Radio Systems